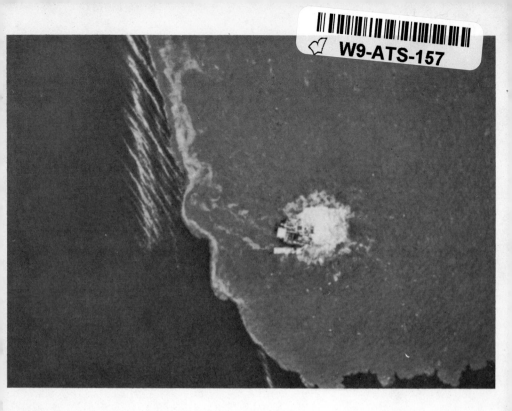

Santa Barbara oil spill, Courtesy of U.S. Dept. of the Interior

It is Estimated that the Printing of this Book on
100 PER CENT *Recycled* Paper has Saved 136 Trees.

AMOS TURK, Ph.D.—Professor, Department of Chemistry,
The City College of the City University of New York

JONATHAN TURK, Ph.D.

JANET T. WITTES, Ph.D.—Department of Statistics,
George Washington University, Washington, D.C.

Ecology
Pollution
Environment

W. B. SAUNDERS COMPANY—Philadelphia—London—Toronto

W. B. Saunders Company: West Washington Square
Philadelphia, Pa. 19105

12 Dyott Street
London, WC1A 1DB

833 Oxford Street
Toronto 18, Ontario

Ecology, Pollution, Environment   ISBN 0-7216-8925-6

Print No.:   9   8   7   6   5   4   3

To Regina,
who is closely related to all of us

PREFACE

Science has been aptly defined as "the body of knowledge obtained by methods based upon observation."* However, the separation of science into categories, or disciplines, such as biology, chemistry, and physics is not based on any natural divisions in such knowledge, but rather on the behavior of scientists. Scientists act according to the demands of their sponsors, whether they be princes or research-granting agencies; to the availability of previously developed research methods and equipment; to what they learn from their teachers, their colleagues, and their pupils; and to established scientific fashions.

The current general concern with problems of the environment is generating pressure to modify some of these "established fashions" so as to create a discipline of environmental science. Evidence of this pressure is seen in the titles of such scientific journals as *Environmental Science and Technology* (American Chemical Society, Washington, D.C.) and *Atmospheric Environment* (Pergamon Press, Oxford, England); such governmental bodies as the Environmental Protection Agency (U.S. Government) and the Department of Environmental Protection (State of New Jersey); and such industrial positions as "Manager of Environmental Sciences" (Borg-Warner Corporation, Des Plaines, Illinois) and "Director of Environmental Control" (Owens-Corning Fiberglas Corporation, Toledo, Ohio).

What does environmental science include? How is it related to the more traditionally established disciplines? A careful observation of the activities of environmental scientists reveals that they

---

*Robert B. Fischer, *Science, Man and Society*. (Philadelphia: W. B. Saunders Co., 1971) p. 4.

are involved with those aspects of biology, chemistry, physics, demography, and engineering that deal with the interactions between man and the rest of the Earth, especially those that tend to disrupt previously established systems and processes. The biological specialty that is concerned with the relations between organisms and their environment is ecology. The impairment of the purity of substances, especially of the components of the biosphere, is pollution. Hence our title: *Ecology, Pollution, Environment.*

This book involves the student in two aspects of environmental science. The first is the actual subject matter of this field. Relevant background material in the physical sciences is presented where it is needed so that previous formal training is not required. We have tried to avoid the temptation to include material that does not relate to this particular study; the book is therefore not to be considered as the basis of a short course in any of the traditional physical or biological sciences, although it does contain some of their subject matter. Instead, it will serve well the purposes of a course in environmental science, as well as supplement a standard course in biology or chemistry. Study problems are offered for each chapter; those that require computations or mathematical reasoning are separated from the others, and numerical answers are provided for them. An annotated bibliography appears at the end of each chapter. The Appendix gives information on units of measurement and on chemical formulas.

The second aspect of environmental science we deal with here concerns the decisions people make about environmental problems. This phase is not presented as "science," but rather as a discussion of social problems and issues. Our purpose is not to provide answers, but to show how various scientific and economic factors must be taken into account so that the final judgments do not lead to unwanted results. This procedure is sometimes called "rationalizing the trade-offs." Such discussions are set alongside the scientific subject matter; they act as a bridge between the acquisition of knowledge and its application.

We thank our many scientific colleagues who reviewed various parts of the manuscript; the staff of the W. B. Saunders Company, who were responsible for editing and production; and Evelyn Manacek, who typed the manuscript.

AMOS TURK

JONATHAN TURK

JANET T. WITTES

▌ CONTENTS

5

AIR POLLUTION 79

6

WATER POLLUTION 109

7

SOLID WASTES 135

8

THE GROWTH OF HUMAN POPULATIONS............... 146

9

THERMAL POLLUTION................................ 175

10

NOISE.. 190

APPENDIX.. 207

INDEX... 211

INTRODUCTION TO ECOLOGY

1.1   WHAT IS ECOLOGY?

There is no machine that can match living organisms in complexity and diversity. Animals and plants, unlike machines, can feed and repair themselves, adjust to new external influences, and reproduce themselves. These abilities depend on very complex interrelationships among the separate parts of the body. Thus, each of us human beings is far more than a sum of brain, heart, liver, stomach, and other organs. What affects one part of the body affects all. Each of us is thus a living **system** of interdependent parts; the system functions in an extremely complex manner, so complex that it is far from fully understood. The study of such functioning is the science of **physiology.**

Even with all the built-in mechanisms of life, however, an individual plant or animal cannot exist as an isolated entity; it is dependent upon its environment. Specifically, it must ingest food, water, and minerals, dispose of wastes, and maintain a favorable temperature. The study of the interactions between living systems and their environment is the science of **ecology.** Taken together, living organisms and the non-living matter with which they interact constitute an **ecosystem.**

1.2   ECOSYSTEMS AND NATURAL BALANCE

We have grown accustomed to the notion that man and nature have cooperated in relative harmony for thousands of years. The technological explosion of the 20th century is blamed for despoiling air and water and for destroying all aspects of environmental calm. But there are many early examples of man's destruction of his environment. Man's disregard for nature is reputed to have contributed to the collapse of the great Central-American Mayan religious centers in about A.D. 800; to the Black Plague, which killed nearly half the population of Europe in the 15th century; and to the devastating Irish potato famine of the 19th century. In order to appreciate the effects of man on the balance of nature and to understand why technology is such a potent threat to the processes of life on Earth, it is vital to understand how ecosystems behave.

When one speaks of a "balanced" person, it is somehow implied that all is well, that he is functioning normally. To be "out of balance" is to be disrupted in some way; the situation is regarded as abnormal and detrimental. The same terminology and thinking is applied to ecosystems. A balanced ecosystem is considered to be in a healthy condition. We must therefore examine the concept of balance more closely.

A balance is an equality of oppositions: a chemical equation is balanced when the opposite sides are equal; a scale is balanced when the weights or forces on each pan are equal; population size is considered to be balanced when the opposing processes, the birth rate and the death rate, are equal. What happens when a balance is disturbed? That depends on the nature of the system. Some systems go out of balance easily; others resist change. As an example of the latter type, your blood is balanced between acidity and alkalinity; that is, opposing chemical processes maintain a constant level (one that is very slightly alkaline). Now suppose a certain quantity of acid is added to your blood. The chemical makeup of blood is such that it will tend to oppose this disturbance of its acid-alkali balance and to return toward its normal condition. This resistance to disturbance of acid-alkali balance is called **buffering**. Buffering is thus a protective action. An organism tends to maintain a balance of various life processes by feeding itself, keeping itself in repair (healing itself), and adjusting to external changes. This tendency to maintain a stable internal environment is called **homeostasis**.

A natural ecosystem, such as a grassland, has many oppositions. Organisms are born and die. Moisture and nutrients travel out of the soil and they are transferred back into the soil. Furthermore, many of these oppositions are exquisitely protected against disruption. During a dry season, when the mice in a grassland have less food and their birth rate decreases, they go back into their burrows and hibernate so that their death rate also decreases. Thus, their behavior protects their population balance as well as that of the grasses, which are not consumed by hibernating mice. Such a tendency is called **ecosystem homeostasis.**

The "balance of nature" is thus an expression that refers to the condition of natural ecosystems that maintains their existence by appropriate opposition of processes and by regulatory mechanisms that protect these processes against disruptions.

1.3 FOOD CHAINS

The earth, in its travel through space, exchanges very little matter with the rest of the universe. Thus, for all practical purposes, life started on this planet with a fixed supply of raw materials. There are finite quantities of various elements — for example, oxygen, nitrogen, carbon, hydrogen, and iron. The chemical form and the physical location of each element can be changed, but the quantity cannot.* In contrast, the earth is continuously receiving energy from the sun, and will continue to do so far enough into the future so that we need not concern ourselves about when it will cease. There are, however, two limitations on the use of the sun's energy. First, much of it is used simply to heat the earth to a temperature at which life can exist. Second, the amount of energy received per day is limited. However, as we shall see, energy can be stored on earth in various forms and used at a later time.

Both plants and animals need raw materials (matter) and energy to live. Plants use simple materials and energy from the sun to synthesize organic compounds which store energy. Animals then consume these plant products as a source of both matter and energy.

Let us examine these processes in more detail. Plants are characterized by the ability to convert the raw materials carbon dioxide and water to organic compounds with the aid of energy obtained from sunlight. This process is known as **photosynthesis.** The important thing to realize here is that the raw materials used are poor in stored energy while the products of photosynthesis are rich in

---

*This statement does not apply to radioactive elements. This topic is discussed separately in Chapter 4.

energy.* We can understand the energy content of substances in terms of their ability to produce heat when they burn in air. An energy-rich material is one that can be burned with the production of heat, while an energy-poor compound cannot. Thus, cellulose (wood), fat, sugar, coal, and petroleum are energy-rich substances, whereas carbon dioxide, water, nitrogen, and granite are energy-poor.

Now let us imagine that plants could transfer energy-poor compounds into energy-rich compounds indefinitely, and that there were no other types of changes occurring on Earth. Eventually, the energy-poor compounds such as carbon dioxide and water would be depleted, there would be a huge accumulation of energy-rich substances, such as those comprising leaves, seeds, and twigs, and life would cease. Of course, this has not occurred, because raw materials cycle. All organisms produce products. First, there are waste products associated with the act of living. Plants produce more oxygen than their life processes require; animals exhale carbon dioxide. However, organisms themselves can be viewed as products. A tree produces leaves, twigs, a trunk; a lion produces bones, a mane, a tail. All these parts of an organism, indeed the entire organism, when it is either living or dead, is a product which can be consumed by other organisms.

A generalized cycle can be visualized in the following manner: Suppose substance A is a vital raw material for a certain species of organism #1, and substance B is one of its products. Substance B

---

*The energy we are considering is related to the processes that occur in living matter. We are not concerned here with nuclear energy, which is taken up in Chapter 4.

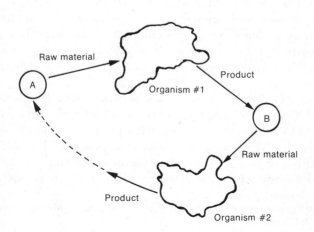

must then be a vital raw material for a second organism, which then produces a new product. This alternation of products and the organisms that consume them must continue until the cycle is complete; that is, until some organism produces our original substance A. When averaged over a long period of time, each product must be produced and consumed at the same rate; consumption must balance production. If any substance in the cycle, for example, substance B, were produced faster than it was consumed, it would begin to accumulate. The population of organism #2 would then increase in response to its greater food supply, and it would begin to consume more of substance B. Eventually the consumption of B would be increased enough to match its production, and the cycle would thus have regulated itself. Conversely, if B were produced at a rate *less* than the demand for it, the population of organism #2 would decline because of this insufficiency. The resulting decreased demand would enable production to catch up with consumption, and balance would be reestablished. Any permanent interruption of the cycle, for example by extinction of one of the species of organisms and the failure of any other organism to fulfill its function, would necessarily lead to the death of the entire system.

Generally, the larger the system, the greater will be its ability to adjust to inequality. In very small ecosystems, balance must be nearly exact if the system is to survive. A simple example will serve as a paradigm for viable ecosystems. Suppose you place some pond water, which contains microorganisms, into each of five large jars. You then add the following:

Jar No. 1: A few snails and no algae.

Jar No. 2: No snails and some algae.

Jar No. 3: A few snails and some algae.

Jar No. 4: A few snails and a lot of algae.

Jar No. 5: Many snails and a small amount of algae.

Then you seal all five jars so that no matter (including air) can enter or leave. All jars are exposed to sunlight (see Fig. 1.1). What will happen to each jar?

In the first jar, the snails would simply suffocate, just as a human being would suffocate in a sealed room. Animals use oxygen as one of the raw materials necessary for life. When the oxygen is depleted, the animals die.

In the second jar, the algae would suffocate, for plants use carbon dioxide as one of the raw materials necessary for their life, and after a while all the available carbon dioxide would be used up.

The third jar is most likely to become a balanced, stable ecological system. The algae use water, carbon dioxide, and sunlight

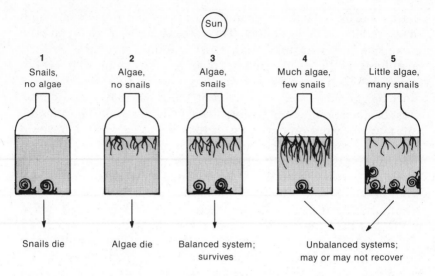

(Snails are animals; algae are plants)

Figure 1.1. Snail-algae ecosystems.

to make sugars. Oxygen is eliminated as a waste product. Since the algae originally introduced were living, healthy plants, they contained quantities of stored nitrogen, sulfur, phosphorus, and various other elements. These materials would be used for the synthesis of proteins, vitamins, and other necessary compounds.

Plants produce more food than they need. This overproduction allows animals to eat parts of the plants, thus obtaining energy-rich sugars and other compounds. These compounds, in turn, are broken down during respiration, a process that uses oxygen and releases carbon dioxide. Other animal waste products, such as urine, though relatively energy-poor, are consumed by micoorganisms in the pond water. The waste products of the microorganisms are even simpler molecules which can be used by the algae as a source of raw materials. The consumption of dead tissue by aquatic microorganisms is essential both as a means of recycling raw materials and for waste disposal. The interaction among the various living species permits the community of the jar to survive indefinitely. In fact, biology laboratories have sealed aquariums in which life has survived for a decade or more. We would like to stress that the picture presented of this cycle has been greatly simplified. The various exchanges of materials that occur in such a simple system are actually very complex and much beyond the scope of this book to trace in detail. Cycles which occur in nature are so complicated that they are not fully understood today.

Now let us consider the fourth jar, which had very few snails and a large quantity of algae. These animals would be healthy at first, for they would have plenty to eat. Also, because of the abundant food supply, the snail population would increase. At the same time, because the carbon dioxide concentration would be low, plant growth would be stunted and for a while the algae would be eaten faster than it was produced. This situation is inherently unstable, for soon the food supply would be so low that animal growth would decrease. Eventually a balance might be reached. It is important to stress several factors at this point. First, natural systems can regain balance if some outside influence disturbs them. However, the return to balanced conditions takes time, and if the disturbance is too great, the system may never recover. Suppose, for example, that the excess of algae were very great and the available air supply small. The plants might suffocate, in turn causing the death of the animals.

On the other hand, if there were many snails and a small amount of algae (as in Jar No. 5) the snails might eat all the algae on the first day, and the whole ecosystem would die.

The experiment described above outlines a deceptively simple food cycle. There are three elements in this cycle; plants, plant-eating animals (**herbivores**), and decay organisms. The decay products are returned as raw materials to the plant, thus completing the cycle. In nature, **food cycles** (often called **food chains**) are much more complex. In all living systems energy is harnessed by the process of photosynthesis occurring in plant tissues both on land, in grasses and trees, and in water, in seaweed, algae, and the microscopic phytoplankton. The second step in the cycle (or link in the chain) is occupied by the herbivores. On land there are large, hoofed animals, rodents, birds, reptiles, worms, fungi, and insects, as well as many other species. In the water there are various sponges and mollusks, fish of many sizes, and the tiny zooplankton that feed on the phytoplankton. The herbivores are not the end of the chain, for they are consumed by organisms called **carnivores**. Thus, mountain lion eat deer, hawks eat rodents, and many kinds of animals eat insects. A fourth level in the food chain is taken up by the carnivores that eat carnivores, such as trout that eat carp, foxes that eat owls, and birds that eat predator insects. Some animals, like man, eat both plants and other animals. Such species are called **omnivores**. The process of decay adds to the staggering complexity of a food cycle. Millions of different types of organisms

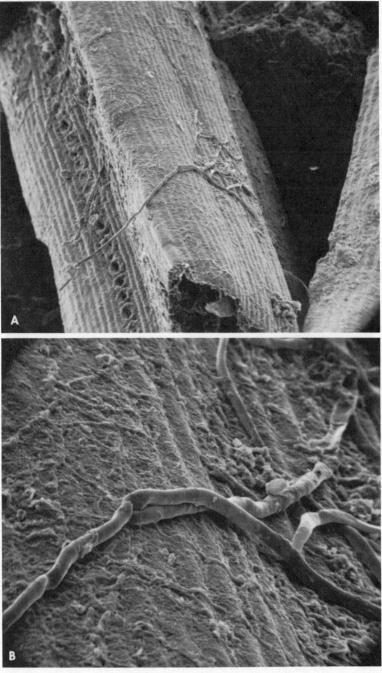

Figure 1.2. See opposite page for legend.

are responsible for decay; some species eat the decay species; decay species eat each other; large decay organisms die and are decayed in turn. No one has unraveled all the details.

Most ecological classifications lump the decay organisms into a single category. With the true degree of complexity thus hidden from view, the plants are called primary producers, the herbivores are called primary consumers, the carnivores that eat the herbivores are called secondary consumers, and so on. A generalized land-based food cycle is shown in Figure 1.3, an aquatic food cycle in Figure 1.4. Some animals are in both cycles. Gulls, for instance, eat fish but are eaten by land carnivores.

It is of interest to consider the efficiency of energy transfer along the food chain.* Let us take as an example a plant that receives 1000 calories of light energy from the sun in a given day. Experiments have shown that only about 10 calories are used to produce plant material. What happens to the other 990 calories? Some of it is used by the plant itself to maintain life. It is common knowledge that a person must eat even if he wishes to remain at a constant weight. This energy is used to maintain proper body functioning. Plants, too, need energy to maintain their metabolism. Some of the chemicals produced by photosynthesis are used to this end. In addition, the process of photosynthesis is not 100 per cent efficient, and therefore a portion of the 1000 calories received is simply lost to the surrounding air as heat. Now, suppose a herbivore, say a cow, were to eat that particular plant. She would receive 10 calories. However, the cow must maintain her own metabolism as well as retain some energy for muscle action to move around to find more grass. The cow's net production—that is, her usable energy gain—would be only about one calorie. Thus, for every 10 calories eaten by a herbivore, only one is available for the carnivore. Tertiary consumers, the carnivores that eat the other carnivores, receive even less. Of the original 1000 calories, they can use only one-tenth of a calorie for weight gain. The energy advantage of the herbivores is one important reason why there are so many more herbivores than carnivores. It is obvious that man, who can occupy

---

*See Appendix I for a discussion of units of energy.

Figure 1.2. "Decomposer" in action. Scanning electronmicrograph of a fungal mycelium on a pine needle being decomposed in the forest floor litter. The upper photo shows a section of the pine needles at 100 × magnification, while the lower photo shows a close-up of the branching fungus at 500 × magnification. (Photos by Dr. Robert Todd, Institute of Ecology, University of Georgia. From Odum: *Fundamentals of Ecology.* 3rd ed. Philadelphia: W. B. Saunders Co., 1971.)

Figure 1.3. Generalized land-based food cycle.

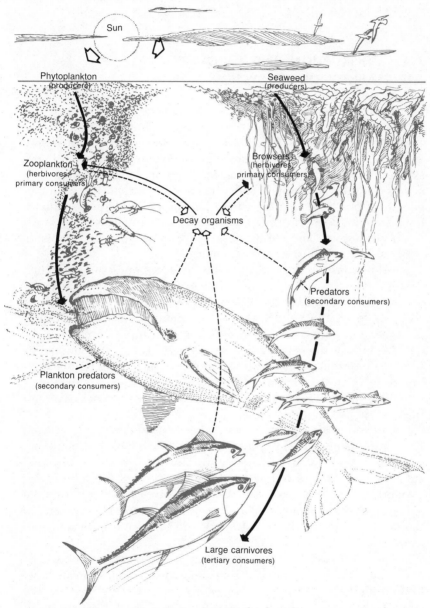

Sun

Phytoplankton
(producers)

Seaweed
(producers)

Zooplankton
(herbivores
primary consumers)

Browsers
(herbivores
primary consumers)

Decay organisms

Predators
(secondary consumers)

Plankton predators
(secondary consumers)

Large carnivores
(tertiary consumers)

Figure 1.4. Aquatic food cycle.

primary, secondary, or tertiary positions in the food chain, uses the sun's energy most efficiently when he is a primary consumer, that is, when he eats plants.

1.4   DISRUPTION AND RECOVERY—
THE STABILITY OF ECOSYSTEMS

Ecosystems are not necessarily in balance all the time, but if they are imbalanced in one direction, they must become imbalanced in the opposite direction at some time in the future if they are to survive with no drastic change in character. In fact, all ecosystems naturally fluctuate. Because we are no longer dealing with a closed bottle, additional factors become important. For example, climate varies from year to year. Other disruptions, such as migration, drought, flood, fire, or unseasonable frost can cause imbalance in an ecosystem. The ability of an ecosystem to survive depends on its ability to adjust to an imbalance.

Let us consider as an example the relative populations of tigers, grazing animals, and grass in a valley in Nepal. Assume that the mountains surrounding the valley are so high that no animals can enter or leave the valley. One year rainfall is limited and mild drought conditions exist. Because water is naturally stored in pools, the drinking water supply is adequate. However, the grasslands suffer from the lack of rainfall. Therefore food is scare for the grazing animals, and many are hungry and weak. Such a situation is actually beneficial for the tigers because hunting becomes easy, and the tiger population thrives.

The following spring, when rains finally arrive, the grazing population is low. This permits the grass to grow back strongly because not much of it will be eaten. However, because the food supply for the tigers is low (and the grazers that are left are the strongest of the original herd) the tigers suffer a difficult time the summer after the drought. The third year there are fewer tigers, so the herd can resume its full strength and balance can be achieved again.

Stable ecosystems, if examined superficially, do not seem to go out of balance at all. Actually, their homeostatic mechanisms function so well that slight imbalances are corrected before they become severe. Such sensitive responses are, in fact, the essence of stability, because if severe imbalances are prevented then it is very unlikely that the system will be destroyed and, as a result, it lasts for a long time. Good examples of stable ecosystems must

Figure 1.5. Moist coniferous forest, a stable ecosystem, in the Olympic National Forest, Washington. (U. S. Forest Service photo. From Odum: *Fundamentals of Ecology.* 3rd ed. Philadelphia: W. B. Saunders Co., 1971.)

therefore be sought in old places. A wonderful example is a redwood forest. To the hiker entering a redwood forest on a hot summer day, two characteristics of the environment are immediately apparent. It is cool, and there is a thick floor of spongy matter. The coolness, caused by the extensive shade of the tall trees, reduces water loss due to evaporation. The spongy floor provides an excellent means of retaining water, much as a sponge would take longer to dry out than the surface of a rock. Thus, the effect of a drought does not have to be balanced by the death of trees. Instead, the moisture in the forest floor serves as a reservoir from which water can be drawn as needed. The coolness of the forest offers still another advantage. Clouds passing overhead are more likely to discharge their moisture where it is cool, for cold enhances the tendency for water to agglomerate into raindrops. Thus the redwood forest tends to cause rain, to restrict evaporation, and to retain a substantial supply of water. Furthermore, decay organisms require moist environments to be effective, so the redwood forest insures the proper recycling of materials by environmental control.

This unusually stable ecosystem also exhibits other highly effective homeostatic abilities. One of these can be observed readily by walking through the forest on the west side of Route 1 as it runs through the town of Big Sur, California. Some time ago a tremendous fire swept through the forest. However, because the bark and the wood of the redwood tree are fire-resistant, the big trees were scarred but not killed. Because the forest environment was maintained the forest community was quickly regenerated. If a fire of equal intensity had burned through a stand of pine, the trees would have died and many more years would have been needed for the forest to be rebuilt.

We have defined a stable ecosystem as one that is difficult to upset because its regulatory mechanisms are highly effective. Now we ask the question, "What factors favor stability?" One clue comes from the observation that natural ecosystems are generally more stable than artificial ones. It is notable that in all stable natural ecosystems both the quantity of organic matter and the number and diversity of species are very large. The role of the quantity of organic matter in stabilizing the system is incompletely understood. It is clear that quantities of decaying vegetation serve to absorb and retain water. Decaying vegetation also provides a reservoir of minerals which are released continuously by decay organisms. In addition, there is some evidence that incompletely decayed chemicals dissolved in the water act as environmental hormones just as chemicals in the blood of animals act as hormones for the regulation of homeostasis. For example, when certain species of algae become very dense in one area, chemicals which inhibit their growth are excreted into the water. It is also possible for different species to work together to control the environment. For example, certain molds produce antibiotics which kill bacteria. This control of the bacterial population in turn assures survival of the animal population and helps maintain the balance of the entire ecosystem.

There are, of course, many different species of plants and animals on Earth. No one knows how many, but we do know that the number is very large. There are over 600,000 known species of insects alone. Each species performs certain functions and occupies certain locations, or habitats. The combination of function and habitat is called an **ecological niche**. Naturally, the niches occupied by different species are not all exclusive; this overlapping produces competition. Natural ecosystems are characterized by the fact that they contain many species, and that many of the niches occupied by these species are very closely related. One study revealed that in a certain forest there were 60 species of mites (a small animal which eats fallen plant litter). All ate approximately the same food, and, although one species was predominant, all the other species seemed viable. Thus, there seems to be a relationship between the stability of an ecosystem and the presence of many species occupying overlapping niches. Why do such circumstances favor stability?

The answer can be illustrated by a simple example. Home gardeners are familiar with the fact that one can buy lawn seed that germinates best in sunny areas or, if they wish, seed that sprouts in the shade. If a mixture is planted in the sun, more of the sunny seed grows than the shady, although some shady seed survives. Now let us say that the lawn is left unmowed, the grass is allowed to reseed itself, and a tree starts to grow. As more shade develops, the majority of the population shifts from mostly sun-loving grass to mostly shade-loving grass. To the casual observer no change has taken place; a healthy grass existed before the tree was present and continued to exist after the tree had grown. However, if only sunny lawn seed (one species) had been planted, the emergence of a tree would have destroyed the grass population.

Another example that may be cited is that of the Eastern Range of the Colorado Rockies, where the two sides of a canyon often exhibit two separate ecosystems, even though they are each at the same altitude and may be only a half-mile apart. Thus, the ponderosa pine, juniper, and small cacti will predominate on the sunny side, and the blue spruce, Douglas fir, and flowering plants will predominate on the shady side. Although both sides of the canyon contain representatives of all species, the niche of the pine is sufficiently different from that of the spruce to preserve the separation. If a prolonged drought were to strike the canyon, there would probably be some shift in the plant population in favor of those species that are more effective in controlling water. Thus, changing conditions often change the order of dominance of species. The advantage of diversity is that the changes that do occur are relatively mild and the ecosystem is not disrupted.

One might ask, however, how diverse populations can exist without a few species dominating so completely that they destroy the others. The answer seems to be that species occupying closely related niches have evolved in such a way that conflict between them is minimized, or at least kept down to a level where they can all survive. Consider another example from the Colorado Rockies. Two predators, the fox and the coyote, live in the same area. The fox is smaller and stealthier, while the coyote is larger, stronger, and faster. Both hunt rabbits and large rodents. In addition, the fox, but generally not the coyote, kills birds and small rodents. On the other hand, coyotes, hunting in groups, can capture small deer. Thus, by having slightly different abilities, the fox and the coyote occupy slightly different ecological niches, and both survive.

We have emphasized that the various species within an ecosystem are interdependent. How near to each other need dependent species be? The answer is that the range of distances varies considerably. An example of very close proximity between interdependent species is rock lichen. This lichen grows on bare rock and resembles an extremely thin layer of vegetation. Actually, the lichen is a mixture of a fungus and an alga. The fungus, which does not contain chlorophyll and thus cannot produce its own food by photosynthesis, obtains all of its food energy from the alga. In turn, the alga cannot retain water and would dehydrate and die if it were not surrounded by fungus. Here the dependence is direct because the organisms must grow together in order to survive. Such a relationship is called **symbiosis**.

On the other hand, some species may depend on others living within distances of several miles. For instance, in a forest there is abundant plant life as well as both herbivores and predators. In a stable forest ecosystem, the trees provide shelter, shade, moisture, and temperature control. Herbivores use the trees and other plants for food. In turn, the predators eat the herbivores. Fungi and bacteria perform the necessary decomposition of wastes. We know from our earlier discussion of food cycles that the system needs both plants and herbivores to continue, but what function do the predators contribute to the system? Predators help maintain population control of various species. If there were no predators, we would expect the size of the herbivore population to display dramatic cyclical changes. It would grow rapidly until the population density was too large for the available food supply. Then there would be mass starvation and disease, causing a rapid population decline. Once the herbivore population size was low, food supply would again be abundant and the cycle of growth and decline could begin anew. However, such a system is unlikely to continue for many cycles, because when the population of the herbivores is very low, extinction becomes a significant threat. Thus the predators, by limiting the population size of the herbivores, tend to prevent overpopulation which leads to eventual starvation and possible extinction.*

In large continental areas entire ecosystems depend on other ecosystems which may be located thousands of miles away. The

---

*In Chapter 8 we shall see how these same principles are of use in studying human population size.

life around a large river depends on the yearly cycles of river flow. In turn, the river flow depends on the water balance at the tributaries. This water balance is controlled by the forest systems. Thus the overall state of the forests on banks of a small creek has a direct effect on the life cycles of organisms at the mouth of the river.

There is even some evidence that the stability of a large ecosystem in one part of the earth might be vital to the stability of the rest of the world's ecosystem. There have recently been discussions concerning the building of a huge dam across the Amazon river. The power produced would be of enormous technological benefit to the peoples of South America. However, the system would flood and hence destroy the Amazon river basin jungle. This jungle, one of the largest and densest forests on the Earth, is the site of a great deal of photosynthesis. It has been calculated that if the forest were flooded, the carbon dioxide concentration in the atmosphere would rise noticeably because the amount of photosynthesis occurring on the face of the Earth would decrease markedly. Since carbon dioxide has the ability to absorb heat from the sun, it is possible that if the Amazon were dammed the additional CO_2 concentration would absorb enough extra heat from the sun to melt the polar icecaps.* That would produce enough water, in turn, to flood most of the coastal land areas (and hence most of the populated areas) of the Earth.

1.5 NATURAL SUCCESSION

We must consider the time spans over which ecological changes occur. We have already shown that a drought will first affect the grasslands, second, the herds of herbivores, and third, perhaps a year later, the predators. However, the effects of a flood or a fire can be immediate. It would be helpful to follow the life of a pond to illustrate both short- and long-term effects. A small pond is often used as an example of a stable ecological system. Animals, microorganisms, and plants all exist in a manner analogous to the experimental jars discussed previously. Temporary imbalances occur and are adjusted as in all natural systems. However, for most ponds the incoming streams and rivers bring more mud into the environment than is removed by the outgoing streams. The effect is small, it may not be noticed in the lifetime of one man, but it does exist. A short-term study of the ecology of the pond

---

*See Chapter 5 for a discussion of this phenomenon.

would conclude that the ecosystem was in balance. But mass balance is disrupted by the steady addition of solid matter from incoming streams. In time, the lake will begin to fill up with mud. The vegetable and animal life will change. New plants that can root in the bottom mud and extend to the surface where light is available will appear. The trout give way to carp and catfish. If an ecologist studies the lake at this stage he might again say that it was a stable system. Again, minor imbalances and adjustments could be observed but the overall system appears stable.* Yet mud continues to flow into the lake. In addition, since it is common for the plants in the lake at this stage to produce more food than is consumed by the herbivores, the bottom fills up with humus. Eventually the lake may become so shallow that marsh grass can grow. Once again, a marsh is considered a stable system over a short range of observation, but in most cases it is gradually evolving into a meadow. If the climate is right, trees can start to grow. First, shrubs appear, then quick-growing soft woods like soft birch, poplar, or aspen. The soft woods are replaced by pine; and finally, in what is called the climax, the pine is replaced by hardwoods.

During the succession of plant species, each species prepares the way for the next but contributes to its own extinction. The marsh grass could never grow without the rich soil of the partially decayed seaweed. However, by producing the rich soil, the seaweed helps to fill the lake and thereby destroy its own environment.

We cannot estimate the amount of time typically required to fill a lake, because that depends on many factors, such as the original volume of the lake and the net rate of accumulation of solids. However, the progression from grassland to a climax forest has been measured. In the southwestern United States, grasslands give way to shrub thickets in one to 10 years, the shrubs become pine forests in 10 to 25 years, and the pine forests give way to hardwoods after about 100 years. We emphasize again that for an ecosystem to be indefinitely stable there must be a complete balance. This is never the case on Earth. Daily, seasonal, yearly, and long-term fluctuations occur in all natural systems, even in the most stable ones. The ultimate consideration of concern for the continuation of life on this planet is that the sum total of all the changes results in worldwide total balance.

The change from pond to meadow to forest, described above, is an example of **natural succession**. Succession is defined as the sequence of changes through which an ecosystem passes as time

---

*The acceleration of these processes is one of the effects of water pollution discussed in Chapter 6.

goes on. The **climax** is the "final" stage, the stage that is "unchanging." Of course, the word final is used with reservation, because the slow process of evolution changes everything. The composition of the climax depends on temperature, altitude, seasonal changes, and patterns of rainfall and sunlight. Descriptions of some of the major climax ecosystems of North America follow:

Forests that consist largely of deciduous trees need adequate moisture and a relatively temperate climate. Therefore, they do not exist farther north than central Maine or southwestern Quebec or farther south than central Georgia or southern Louisiana. Rainfall must be between 30 and 60 inches per year and the average temperature during the growing season should be about 60° to 65° F.

If the average temperature and rainfall levels are lower and the winters longer, as they are in parts of Maine, Alaska, Minnesota, and Canada as well as in the mountain regions of the West, the climax forests are composed largely of various species of evergreen.

In still colder regions the climax consists of vegetation almost exclusive of trees. This climax ecosystem, known as the **tundra**, consists of plants whose growing season is two months or less and which can survive severe conditions of wind and frost (see Fig. 1.6). The growth is all low, dense, quick to flower, and, we might add, quite beautiful.

In temperate areas in which the rainfall is too low to support forests, the climax ecosystem consists of prairie (see Fig. 1.7). In the driest areas (less than 10 inches of rainfall per year) the climax is a desert, either barren or able to support scrub brush and cactus (see Fig. 1.8).

Over geological time the cycling of raw materials has been quite effective; otherwise, life could not have existed for these past millions of years. However, it is evident that the cycles are not complete. For example, the accumulation of large deposits of coal and petroleum is evidence of the fact that raw materials have been removed from the cycle. The growing salinity of the oceans is another such example. Perhaps, in time, geological upheavals will serve to redistribute these materials.

1.6   THE ROLE OF MAN

Man is, of course, a part of the ecosystem of Earth. His survival depends on the survival of hundreds of thousands of species of plants and animals. Why, then, do we consider man separately? The answer is that today man is drastically altering the ecosystems

Figure 1.6. Tundra in July on the coastal plain near the Arctic Research Laboratory, Point Barrow, Alaska. (Photos by the late Royal E. Shanks, E. E. Clebsch, and John Koranda. From Odum: *Fundamentals of Ecology.* 3rd ed. Philadelphia: W. B. Saunders Co., 1971.)

Figure 1.7. Natural temperate grassland (prairie) in central North America. A. Lightly grazed grassland in the Red Rock Lakes National Wildlife Refuge, Montana, with small herd of pronghorn antelope. B. Short-grass grassland, Wainwright National Park, Alberta, Canada, with herd of bison. (From Odum: *Fundamentals of Ecology.* 3rd ed. Philadelphia: W. B. Saunders Co., 1971.)

Figure 1.8. Two types of deserts in western North America. *A.* A low altitude "hot" desert in southern Arizona. *B.* An Arizona desert at a somewhat higher altitude with several kinds of cacti and a greater variety of desert shrubs and small trees. (From Odum: *Fundamentals of Ecology.* 3rd ed. Philadelphia: W. B. Saunders Co., 1971.)

of the Earth and has within his power the ability to destroy many of them totally. Such alteration or destruction may be inadvertent. For instance, when a land developer fills a marsh to build houses, he is destroying the breeding ground of thousands of migrating birds, thus causing considerable change in the many ecosystems to which these species belong. In some instances such destruction is intentional. Spraying a swamp with DDT in order to eradicate malaria-carrying mosquitos is a deliberate, although not necessarily unreasonable, attempt to destroy an element of a natural ecosystem. We have pointed out that ecosystems change in time because of climatic variations, natural succession, and evolution. But these involve long time spans. Important evolutionary changes in the higher animals and plants take millions of years. But man can change the face of the earth in a matter of decades. Before man, evolutionary processes of adaptation have been fast enough to assure survival of life on earth, but we can no longer look to evolution to catch up with the changes produced by man's technology. One example will give an idea of how man has accelerated terrestrial processes: In the past 50 years, Lake Erie has changed to an extent that would have required 15,000 years by the processes of natural succession. It is now considered to be in imminent danger of turning into a huge swamp.

What are the consequences of man's technology to the survival of mankind and all life on earth? This is a very grave and sobering question, for which there is no easy answer. The remaining chapters will consider several facets of man's ecosystem, introduce the reader to some important concepts, and equip him for further study.

PROBLEMS

1. Define ecology; ecosystem; homeostasis.

2. Consider two outdoor swimming pools of the same size, each filled with water to the same level. The first pool has no drain and no supply of running water. The second pool is fed by a continuous supply of running water and has a drain from which water is flowing out at the same rate at which it is being supplied. Which pool is better protected against such disruptions of its water level as might be caused by rainfall or evaporation? What regulatory mechanisms supply such protection?

3. Classify each of the following substances as energy-rich or energy-poor, with specific reference to its ability to serve as a food or a fuel: sand; butter; paper; fur; ice; marble; paraffin wax.

4. What is a food cycle? Sketch a diagram of a food cycle that primarily involves life in the air, such as birds and insects, their predators, and the land or aquatic plants that supply their energy.

5. It is desired to establish a large but isolated area with an adequate supply of plant food, equal numbers of lion and antelope, and no other large animals. The antelope eat only plant matter, the lions, only antelope. Is it possible for the population of the two species to remain approximately equal if we start with equal numbers of each and then leave the system alone? Would you expect the final population ratio to be any different if we started with twice as many antelope? Twice as many lions? Explain your answers. (Assume that lions and antelope have the same body weight.)

6. State and explain three factors that tend to stabilize an ecosystem.

7. Define ecological niche; symbiosis.

8. Explain the role of predators in an ecosystem.

9. How does diversity contribute to the stability of an ecosystem?

10. Define natural succession. What factors bring about changes in an ecosystem? What is the climax of an ecosystem? Cite three examples of a climax ecosystem.

11. Would you say that the Earth includes many ecosystems that are relatively independent of each other, or that it contains only one ecosystem that occupies the entire biosphere, or that both statements are true? Present arguments in favor of your position.

The following question requires arithmetical computation:

12. Assume that a plant converts 1 per cent of the light energy it receives from the sun into plant material, and that an animal stores 10 per cent of the food energy that it eats in its own body. Starting with 10,000 calories of light energy, how much energy is available to a man if he eats corn? If he eats beef? If he eats frogs that eat insects that eat leaves?

ANSWER

12. 100 cal; 10 cal; 1 cal

BIBLIOGRAPHY

The basic textbook on ecology is

Eugene P. Odum. *Fundamentals of Ecology.* 3rd ed. Philadelphia: W. B. Saunders Co., 1971. 574 pp.

A periodical issue devoted in its entirety to "The Biosphere" is

Scientific American. September, 1970. 267 pp.

A very valuable and scholarly book that is oriented toward theoretical biology is

Philip Handler, ed. *Biology and the Future of Man.* New York: Oxford University Press, 1970. 936 pp.

There are several books that are not limited to ecology but are broadly concerned with environmental problems. Three are cited below; the first is oriented toward chemistry, the second and third are more general:

American Chemical Society. *Cleaning Our Environment: The Chemical Basis For Action.* Washington, D.C., 1969. 249 pp.

Paul R. Ehrlich and Anne M. Ehrlich. *Population, Resources, Environment.* San Francisco: W. H. Freeman and Co., 1970. 383 pp.

Ernest Flack and Margaret C. Shipley, eds. *Man and the Quality of his Environment.* Boulder: University of Colorado Press, 1968. 251 pp.

A number of more popular books sound the alarm about the environmental dangers to man's survival. Five of these are

Ron M. Linton. *Terracide.* Boston: Little, Brown and Co., 1970. 376 pp.

J. Rose, ed. *Technological Injury.* New York: Gordon and Breach, Science Publishers, 1969. 224 pp.

Robert Rienow and Leona Train Rienow. *Moment in the Sun.* New York: Ballantine Books, 1967. 365 pp.

Melvina A. Benarde. *Our Precarious Habitat.* New York: W. W. Norton and Co., 1970. 362 pp.

Osborn Segerberg, Jr. *Where Have All the Flowers Fishes Birds Trees Water and Air Gone?* New York: David McKay Co., 1971. 303 pp.

Those readers interested in the cost of environmental control might refer to

Thomas D. Crocker and A. J. Rogers III. *Environmental Economics.* Hinsdale, Ill.: Holt, Rinehart and Winston, The Dryden Press, 1971. 150 pp.

2

AGRICULTURAL ENVIRONMENTS

2.1 AGRICULTURAL ECOSYSTEMS

When man first evolved from his ape-like ancestors he ate meat and various plant products, depending on what he could manage to collect from day to day. He hunted and dragged his food back to his den or was hunted and dragged back to some other predator's lair. He competed with other herbivores for plant foods, and life was very hard during drought, flood, or pestilence.

The environment of the Earth was not appreciably altered by the existence of early man. There were two reasons for this: First, man's early technology was very limited. His stone and wooden tools for digging and hunting were competitive with the tusk of the mammoth and the claw of the tiger, but certainly were not overwhelmingly superior. Second, his numbers were small and, even if he had been technologically advanced, his presence on Earth was too sparse to affect its environment to any significant degree.

But the race between social and biological evolution is no contest. Once man acquired the ability to create any technology, even a stone axe, and to transmit his knowledge from generation to generation, one invention followed upon another and his ascendancy over his competitors became more and more firmly established. His successes made him more numerous; his increased population and his enhanced abilities reinforced each other, and finally the combined effects began to alter the Earth in a noticeable way.

Man's first significant Earth-changing invention was agriculture. (We use the word here in its broadest sense to include both animal husbandry and cultivation of crops.) To understand why agriculture creates a fundamentally new kind of ecosystem, recall from Chapter 1 that a naturally balanced system is self-sufficient and need not exchange any matter, by either giving or taking, with the outside. Consumption and production of nutrients are passed around the food cycle continuously, with no net gains or losses. Such a system may be an isolated area, such as a valley ringed by high mountains, or it may be a sealed jar of snails, algae, and microorganisms.

Now imagine that you eye the sealed snail-algae jar hungrily— you want to eat some snails. You yourself are not part of the jar system; you are outside. So you open the jar and remove some snails. This may satisfy you for a while, but, if you continue, the final choices are inevitable. Either you exhaust the nutrient in the jar and must move on to some other source of food, or you must replenish the jar by a quantity of nutrient equivalent to the snails you have removed. This example is an illustrative model of an agricultural ecosystem as opposed to a naturally balanced one. We have said that agriculture was a technological invention. It enabled man to extract food from the earth far more efficiently than he ever could by gathering or hunting—so efficiently, in fact, that a given land area could be made to feed more people than just the farmers or shepherds who tended it. Thus were some men freed from the biblical imprecation that "in the sweat of thy face should thou eat bread." They could leave the fields and congregate in cities where they could pursue non–food-gathering activities, such as arts and technology, including the further development of devices to improve agriculture. But men in the city are outside the system that produces their food; they are a drain, like the person who removes snails from the jar. Such a drain interrupts the food cycle and therefore engenders an ecological imbalance. And the more efficient the food-growing technology, the worse is the ecological disruption. Of course, man has understood for many years that his choices were either to refertilize the land or, eventually, to move elsewhere. In some areas, man has been quite successful in keeping farmlands fertile. For example, some regions of Europe have been farmed successfully for thousands of years. During that period of time the soil has even been enriched by fertilization with manure (human and animal). However, in some places, as we will see, man still continues to destroy huge areas of previously fertile land.

We mentioned in Chapter 1 that the diversity of an ecosystem contributes to its stability. For example, the existence of two types of pine trees, one preferring relative dryness and the other, relative wetness, helps insure the survival of the forest even after years of low rainfall. In turn, the presence of the forest maintains the health of the soil, which contains the decay organisms necessary for the decomposition of fallen trees, leaves, branches, and other forest litter. Agriculture by its nature is not so diverse; no farmer wants rabbits in his carrot patch, birds in his corn field, or even carrots in his corn field.

In this chapter we shall outline the traditional methods whereby man has increased the production of his farms, and we shall discuss how these methods have disrupted ecosystems and have sometimes resulted in disaster. Some of the difficult problems inherent in the solution will also be considered.

2.2   THE TECHNIQUES OF AGRICULTURE

Four important agricultural techniques have been used since ancient times to help the farmer to maximize his yield.

(a) Large non-human consumers (mammals, reptiles, and birds) are physically separated from the crops. It is obvious that the harvestable yield of a given area will increase if animals are not allowed to eat the crops. A mature ecosystem is in balance because animals eat the excess produced by plants; when there are no animals, great excess results. If the farmer is the only herbivore in a given field, then he can raise more than he can eat and he has excess to sell. Thus the use of fences to keep herbivores away from crops can increase a harvest yield. The effect on the herbivores is less pleasant. Their freedom to seek new feeding grounds in times of scarcity is curtailed; they have less to eat and many die. For example, deer in North America have been moved into smaller and smaller areas as more forest is cut and more fields are fenced in. (However, man has also diminished the population of the deer's natural predators, so that some of the areas which still remain available to deer are actually overpopulated with them.)

It is a relatively simple task to keep large land animals away from an agricultural area; the effect is to increase the number of humans on the Earth at the expense of other animals. The problem of reducing competition from insects is much more difficult and will be discussed in Chapter 3.

(b) Only those plants which man wants are cultivated. Plants are useful as food if they produce large, edible seeds rich in protein or energy or if they concentrate vitamins or starch in their roots, leaves, or fruits. Some plants, such as tobacco and cotton, are cultivated for purposes other than food. In any case, man can increase his agricultural yield simply by growing highly productive crops in preference to the less productive ones—the weeds. Agricultural techniques enable man's crops to grow with much less competition for water, sunlight, and soil nutrients than they would if they lived in the wild. Weed control, like protection from nonhuman herbivores, is essential to agriculture. Unfortunately, as we will see in later examples, the "weeds" are often part of the balance of nature; their destruction has in the past destroyed the land. The theory of raising only the most productive crops is, of course, sound. But, in practice, weed control by plowing has been directly responsible for the removal of topsoil faster than it can be formed by the natural process of erosion of rocks.

(c) *Land is refertilized.* Artificial recycling or fertilization of farmlands is essential if man is to remove nutrients in the form of grains, fruits, vegetables, and the like from the land. Otherwise the land will lose its fertility.

(d) *Land is irrigated.* The least general of the four techniques of agriculture is irrigation. This technique is necessary only in dry areas where man wishes to grow more food than would ordinarily be possible.

All of these four techniques are necessary to the development of modern agriculture. Each one, when misused, has led to the destruction of farmlands. At present, millions of acres of cropland are being lost each year. Fertile land is presently being destroyed faster than it is being formed. Obviously the situation cannot continue indefinitely. Therefore, let us look at some of man's classic blunders so that we may understand how such losses come about.

2.3   AGRICULTURAL DISRUPTIONS

It is generally known that the agriculture of India cannot quite keep up with the needs of its people and that, consequently, famine is never more than a few dry seasons away. It is less well known that approximately two-thirds of the cropland has been completely or partially destroyed by erosion or soil depletion caused by man. The area around Sind near the mouth of the Indus River is typical of a destroyed cropland. With the exception of areas under irrigation, Sind is now a barren, infertile, semi-desert region. Yet, archeological diggings have uncovered the remnants of highly civilized inhabitants who farmed there over 4000 years ago. Moreover, fossils of the native animals of the area show that elephants, water buffalo, tigers, bears, deer, parrots, wolves, and other similar forest dwellers lived there. The early settlers built temples with fire-baked bricks; undoubtedly, wood fed the fires, and the wood must have come from forests in the nearby areas that are now desert. The details of the transformation from forest to desert are largely conjectural because there are no recorded weather reports over these 4000 years. However, we can suggest a plausible mechanism. Remember that a forest maintains a cooler environment than a grassland. Part of this coolness is due to the water-holding ability of the forest ecosystem. Now, rain clouds passing over a hot tropical steppe will generally rise and precipitation will seldom occur, while rain clouds passing over cool jungle will be induced to drop their moisture. Thus, the change in climate which destroyed the land of Sind was probably caused by deforestation. This poses an interesting dilemma. It has already been stated that a farmer should grow only those crops which produce an efficient quantity of food. Forests are poor producers of food for man. But the killing of the Sind forest ultimately destroyed the land, and dead land produces no food at all. Had the early inhabitants of Sind understood the importance of natural diversity, and had they been able to reach a workable apportionment of forest and cropland, they might have been able to grow enough food without destroying their ecosystem.

In the light of this hypothetical example it is interesting to observe what is happening in India today. As overgrazing destroys the ground cover, goat herders, using 20-foot-long sickles, cut the limbs of the trees for food for their goats. The trees die and the soil blows away. But try to tell a man who has a hungry baby at home to save the trees for future generations!

Another story of land destruction comes from the area of the fertile crescent, the "cradle of civilization." The Tigris-Euphrates valley gave birth to several great civilizations. We know that highly sophisticated systems of letters, mathematics, law, and astronomy originated in this area. Obviously, then, men had the time and energy to educate themselves and to philosophize. We can deduce that the food supply must have been at least adequate. Today much of this region is barren, semi-desert, badly eroded, and desolate. Archeologists dig up ancient irrigation canals, old hoes, and grinding stones in the middle of the desert. What must have happened?

Part of the story starts at the source of the great rivers in the Armenian highlands. The forests were cleared to make way for pastures, vineyards, and wheat fields. But the croplands of man, especially if poorly managed, cannot hold the soil and the moisture year after year as well as natural forests or grasslands. As a result, large water run-offs such as those from the spring rains or melting mountain snows tended to flow down the hillsides rather than soak into the ground. These uncontrolled waters became spring floods. As a protective measure, canals were dug in the valley to drain the fields in the spring and to irrigate them in the summer and fall. But the prosperous agriculture invited competition, and wars swept the lands. Tens of thousands of soldiers marched over, ate the food, and killed, enslaved, or chased away the rural inhabitants. The result was that the canals and croplands were abandoned. The neglected canals turned to marsh, the formation of marshes shrunk the rivers, and the diminished rivers could no longer be used to irrigate other areas of land. The water table sank, and the yields per acre today are less than the yields 4000 years ago.

The story of land destruction can be repeated with monotonous regularity. Ancient Carthage was founded on the shores of the Mediterranean in North Africa amid dry but fertile grasslands. Grain was grown in abundance. Today much of this area has become part of the Sahara Desert. In fact, a large part of the Sahara is a by-product of over-plowing and over-farming fertile land, leading to a depletion of the amount of available moisture. To study this process in more detail we shall examine a great American fiasco — the Dust Bowl.

The early European settlers found millions of acres of virgin (that is, unused by man) land in America. The east coast, where they first arrived, was heavily forested, and the task of clearing land, pulling stumps, and planting crops was arduous. Especially in New England, long winters and rocky hillsides contributed to the difficulty of farming. It was natural that men should be lured by the West, for here, beyond the Mississippi, lay expanses of prairie as far as the eye could see. Deep, rich topsoil and rockless, treeless expanses promised easy plowing, sowing, and reaping. In 1889 the Oklahoma Territory was opened for homesteading. A few weeks later the population of white people there rose from almost nil to close to 60,000. By 1900 there were 390,000 — a people living off the wealth of the soil. In 1924 a thick cloud of dust blew over New York City and into the Atlantic Ocean. This dust had been the topsoil of Oklahoma.

In each of the earlier examples it might be contended that the destruction of the land was really caused by changes in climate rather than by man's mismanagement of the land. Indirect evidence, however, strongly implicates man. For instance, in the areas between the Tigris and Euphrates rivers where the canals were not destroyed, the land remains fertile. Similarly, some areas of North Africa near the Sahara still support trees believed planted by the Romans, and geological evidence indicates relatively constant weather patterns in these areas. However, in the case of the Oklahoma Dust Bowl we *know* that a desert was made by man and that the climate did not change.

Over much of the midwestern United States the rainfall is too low to support a forest system. Instead, true prairie, or grassland, represents the ecological climax there. Grasslands, like forests, owe much of their ecological success to their diversity. Thus, both annual and perennial grasses grow in the prairie. The perennial grasses and some low bushes have deep roots, while the annual plants depend on much shorter and less extensive root systems. During dry years there is so little water that many annuals die. However, the perennials, which use water deep underground, are able to live; in doing so, they hold the soil and protect it from blowing away with the dry summer winds. In years of high rainfall, the annuals sprout quickly, fill in bare spots and, with their extensive surface root systems, prevent soil erosion from water run-off. Survival of both types of grasses is insured by minimization of root competition because the different plants have root systems which reach different depths. In addition, not all species flower at the same time of the year, so that the seasons of maximum growth and consequent maximum water consumption differ.

Before the Homestead Act, the dominant herbivore was the bison. Millions of these animals roamed the plains. Their numbers were kept in check by natural forces. Wolf packs hunted them, but because the bison bulls and cows in their prime were too fast and too strong for a wolf, only the sick and old adults and the slow calves were killed and eaten. When bison were scarce the wolves ate field mice, prairie dogs, and other rodents. Coyotes and foxes also preyed on the small rodents. The ecosystem was stable.

Man's contribution was not particularly far-sighted. He killed the bison to make room for his cattle, then killed the wolves and coyotes to prevent predation of his herds. Moreover, he often permitted his cattle to over-graze. In over-grazed land, the plants, especially the annuals, become so sparse that they cannot reseed themselves. The land itself, therefore, becomes very susceptible to soil erosion during heavy rains. In addition, the water runs off the land instead of seeping in, resulting in a lower water table. Because the perennial plants depend upon the underground water levels, depletion of the water table means death for all prairie grasses. The whole process is further accelerated as the grazing cattle pack the earth down with their hooves and block the natural seepage of air and water through the soil.

Man's introduction of the plow to the prairie had an even more severe effect. He dug up and turned over the prairie soil to make room for his crops. We have seen how diversity makes the prairie stable. A commercial farm, however, grows large fields of a single crop, thus destroying the naturally balanced system. The first step in turning a prairie into a farm is to plow the soil in preparation for seeding. At this point, of course, the soil is vulnerable, since the perennial grasses which normally hold the soil during drought have already been killed.

If the spring rains fail to arrive, then the new seeds won't grow and the soil will dry up and blow away. As an emergency measure during droughts, farmers often practice "dust-mulching." By chopping a few inches of the surface soil into fine granules a thin layer of dust is formed. This dust layer aids the capillary action whereby underground water is brought to the surface. Thus the soil remains moist just below the dust, and the seeds sprout. However, before labeling dust-mulching a success, one must examine the total picture. Fertile soil is more than pulverized rock; it consists of decay organisms and partially decayed organic matter. Moreover, prairie topsoils have evolved a balancing mechanism whereby the concentration of decayed organic matter remains approximately constant. Dust mulching brings to the surface much of the previously underground organic material. Once on the surface this organic matter reacts with the oxygen of the air (oxidizes) much more rapidly than it would have had it remained where it was. The result is a decrease both in the concentration of organic matter in the soil and in the soil's fertility. As these losses continue, a time is finally reached when the soil is so barren that it cannot support even the growth of the hardy prairie grasses. Of course, the farmer can replace lost organic matter by spreading manure or other fertilizer, but the fact is that millions of acres of previously fertile farmland have been ruined because the farmers have not re-fertilized adequately. (See Fig. 2.1.)

On the other hand, if, after plowing, the spring rains are too heavy, then the soil may easily wash away before the seeds have an opportunity to grow. Even after seeds have sprouted, the practice of pulling weeds between the rows leaves some soil susceptible to erosion by heavy rains.

From a knowledge of these facts it is easy to understand what

A

B

Figure 2.1. A. Poor agricultural land use that resulted in massive erosion, abandoned homes, and poverty. B. Farmland in Mississippi ruined by soil erosion. (U. S. Forest Service photos. From Odum: *Fundamentals of Ecology.* 3rd ed. Philadelphia: W. B. Saunders Co., 1971.)

happened in Oklahoma. Over a period of 20 to 35 years the soil fertility slowly decreased. Incomplete refertilization and loss due to wind and water erosion took their toll. Finally, when a prolonged drought struck, the seeds failed to sprout and a summer wind blew the topsoil over a thousand miles eastward into the Atlantic Ocean.

The droughts that killed the Oklahoma farms had no lasting effect on those prairies left untouched by man. In fact, these virgin lands are still fertile. In a few thousand years, perhaps, the wind-scarred Dust Bowl will regain its full fertility. We say "perhaps" because similar destruction of the North African prairie left the land so barren that nothing was left to hold the rain that did fall. Two thousand years after the farms failed one can stand in the center of the ruins of a wealthy country estate and watch the sands of the great Sahara blow by.

One might think that it would be impossible to destroy land by importing water (irrigating). Unfortunately, irrigation, too, can be destructive. When rain water falls on mountain sides it collects in small streams above and below ground and it filters over, under, and through the rock formations. In the process of flowing into a large river, the water dissolves various mineral salts present in the mountain rock and soil. Usually these salts are concentrated in the oceans. However, if the river water is used for irrigation, man is bringing salty water to his farm. When water evaporates the salt is left behind and, over the years, the salt content of the soil increases. Because most plants cannot grow in salty soil, the fertility of the land decreases. In Pakistan an increase in salinity decreased soil fertility alarmingly after a hundred years of irrigation. In parts of what is now the Syrian desert, archeologists have uncovered ruins of rich farming cultures. However, the land adjacent to the ancient irrigation canals is now too salty to support plant growth.

In addition, extensive, poorly managed irrigation has been known to raise the water table of the irrigated land. If the water table is too high, plant roots will become immersed in water and will die from lack of air. In the earlier 1960's waterlogging and salinity problems were causing the loss of 60,000 acres of crop land in Pakistan alone.

Occasionally, even an apparently well-planned upset of the balance of nature has produced disastrous results. For thousands of years the peasants of Egypt farmed the Nile Valley. Every spring the river flooded and brought with it water and fresh soil from the Abyssinian Mountains. When the water level subsided the farmers grew what they could during the long hot summer. To increase the agricultural yield for a growing population, engineers had long considered the feasibility of damming the Nile, thus storing the excess water during flood seasons for more efficient use during the summer. The Aswan High Dam, built over a period of 11 years and inaugurated in 1971, was to make this objective possible as well as to provide abundant hydroelectric power. Water storage started in 1964, when the Nile was diverted into a bypass channel during construction, and power has been produced since 1967. Other benefits obtained have included higher yields of cotton, grain, fruits, and vegetables by irrigation during the summer and by reclamation of previously barren land. But there have been problems. First, the sediment that the flood waters formerly washed out to sea, although useless to farmers, did nourish a rich variety of aquatic life. The absence of this nutrient has resulted in the annual loss of 18,000 tons of sardines. Second, the flood waters previously rinsed away soil salts that otherwise would have accumulated and made the soil less fertile. With the flooding under control, the salts are left in the soil. The consequent rise in soil salinity is a threat to the productivity of the land. Third, the sediment formerly protected the delta land in several ways: it served as an underground sealant to minimize seepage that would otherwise drain off various sweet-water lakes, and it helped strengthen natural sand dikes that protected the coastline against erosion by the powerful currents of the Mediterranean. These buffering actions, too, are gone, as the sediment from the Abyssinian Mountains instead sinks behind the High Dam, and the clear, silt-free waters rush downstream more rapidly than before, eroding the river banks and undermining the foundations of hundreds of bridges that span the Nile. Fourth, the loss of silt previously deposited on the downstream fields is a loss of nutrient, and must be compensated by chemical fertilizers. Fifth, the dry periods between floods, now eliminated by irrigation, used to limit the population of water snails which carry a blood fluke called bilharzia. This parasite spreads easily to man, depositing its larvae under the skin, from whence they go on to invade his intestinal and urinary tracts. The resulting debilitating disease, called **bilharziasis,** follows along the paths of the irrigation canals, infecting about 80 per cent of the people who work in them. How can one balance lost manpower and increased human misery against higher crop yields?

Finally, the High Dam seems to be responsible for a serious loss of water, the one essential substance that was never to be in short supply. The lake behind the dam was to have been filled by 1970; it was not yet half full in 1971. Some limnologists (scientists who study the physical phenomena of lakes) predict that it will not be full for a century or more. Losses are suffered by seepage through porous rock and by evaporation in the hot winds of Upper Egypt. Of course, technology may rescue itself. All the problems cited above are under intensive study; means may be found to

Figure 2.2. Impoundment reservoir on an unprotected watershed that completely silted up in less than six years. (U. S. Forest Service photo. From Odum: *Fundamentals of Ecology.* 3rd ed. Philadelphia: W. B. Saunders Co., 1971.)

control the erosion, the parasites, the seepage, and the salinity, and to make enough power to supply fertilizer plants to manufacture nutrients to replace those formerly carried down as silt from the mountains. There is no law of chemistry or physics that denies the possibility of such technological rescue of technology's disruptions. But thus far we are finding that this goal is much more elusive than we had expected.

2.4   LAND AND PEOPLE

Chapter 1 outlined the structure of natural ecosystems. This chapter has discussed how man, the farmer, has disrupted his land through misuse and poor judgment. However, man has also provided a great abundance of food, and, for many people on Earth, a rich and varied diet. In later chapters we shall see other examples of how man's technology makes him more comfortable and secure, yet threatens to disrupt natural ecosystems to a degree that may undo these benefits.

Man has grown dependent on his technology. Moreover, as we shall see, much of this dependence cannot be reversed. Certainly we cannot abandon agriculture. To do so would be to commit all but a fragment of humanity to death by famine, and probably would lead to extinction of our species. But the fact is that if the present destruction of fertile soil continues the food production ability of our planet will be greatly curtailed, for soil is currently being destroyed faster than it is being replenished.

What are the answers? Technology does provide many bases for hope and even for optimism. Hybrid seeds give rise to grain stalks which produce enhanced yields of grain. Anti-erosion measures practiced in this country are actually reclaiming some exhausted land and minimizing farm erosion in fertile areas. Problems of salinity and waterlogging have been solved in some parts of the world. Man is understanding the reasons for his past failures more completely. The world's food production, has, in fact, been increasing steadily over the years, with only a few minor irregularities. Such advances have led some scientists (often called the "Green Revolutionists") to conclude that the world food problem can be solved satisfactorily.

However, though the earth provides more food now than ever before, people are still hungry, and the total fertile area is decreasing. Furthermore, world food production in 1969 showed no increase over that of the previous year—the first such setback in recent times. What to do about over-grazed fields and hungry people thus continues to be a prime problem for all mankind.

PROBLEMS

1. Explain why agriculture is more disruptive to natural ecosystems than are hunting and gathering of food.

2. If a given land area is farmed for many years, does its productivity necessarily decrease? Justify your answer.

3. What are the major techniques of agriculture? Under what circumstances of use or misuse may each of these methods engender a loss of fertility of the land?

4. Chemical defoliation was introduced as a tactic of war in Southeast Asia in the 1960's, and a study group of the American Association for the Advancement of Science concluded that "it is to be expected that in any future wars of this nature, more extensive use will be made of it (defoliation)." Mangrove trees that grow in the river delta areas can be killed by a single application of defoliant, and many sections of mangrove forest have been rendered barren by this tactic. Outline the ecological factors that will determine whether or not such forests will eventually reestablish themselves. Do you think such reestablishment is certain, doubtful, or hopeless? Defend your answer.

Figure 2.3. The destruction of Indochina. (From *Bulletin of the Atomic Scientists,* May, 1971.)

5. What position with regard to the future of agricultural productivity is taken by the group of agricultural scientists identified as the "Green Revolutionists?" What facts support their position?

6. A father and two sons walk along a plain in India, driving several cows before them. The mother follows, carrying a basket on her head. When a cow defecates, the mother scoops up the droppings into her basket. Later, she plasters these droppings onto a wall to dry. Most of the dried dung is then traded for rice; some is kept by the family for use as a fuel to cook the rice. Outline a possible sequence that might have led to the use of cow dung as a fuel. Does the situation described above represent the lowest condition of poverty of the land, or does it lead to still further disruptions? Explain.

BIBLIOGRAPHY

A delightful book about human nutrition from prehistoric to modern times is

Lloyd B. Jensen. *Man's Foods*. Champaign, Ill.: Garrard Publishing Co., 1953. 278 pp.

The reader who wishes to investigate the more optimistic views on the future of agriculture should refer to

Lester R. Brown. *Seeds of Change: The Green Revolution and Development in the 1970's*. New York: Encyclopaedia Brittanica, Praeger Publishers, 1970. 205 pp.
Willard W. Cochrane. *The World Food Problem: A Guardedly Optimistic View*. New York: Thomas Y. Crowell Co., 1969. 331 pp.

More pessimistic views are set forth in the last series of references given in the bibliography of Chapter 1.

Two recent excellent books on agriculture and food resources are

Kusum Nair. *The Lonely Furrow: Farming in the United States, Japan, and India*. Ann Arbor: University of Michigan Press, 1970. 336 pp.

N. W. Pirie. *Food Resources, Conventional and Novel*. New York: Penguin Books, 1970. 208 pp.

3

▌ PESTICIDES

3.1 COMPETITION FOR FOOD BETWEEN INSECTS AND MAN

For thousands of years, man has regarded insects as pests to be destroyed. The description of the locust plague in the book of Exodus in the Old Testament, for example, attests to a historic dislike and fear of insects. Insect invasions have indeed destroyed the crops of man since agriculture began. One normally thinks of large hoofed animals as the earth's herbivores. However, a much larger portion of plant matter is eaten by insects, rodents, birds, and microorganisms.

Throughout history, man has found it much more difficult to isolate his crops from the small herbivores than from the large ones. Very simply, if a cow wanders into a field, the farmer's son can chase her away. An insect, a bacterium, the spore of a fungus, or a locust's egg sac is more difficult to deal with. Because most small species reproduce in great numbers, their total eating capacity is very great. In addition, insects have been serious pests because they act as carriers of disease organisms. Malaria and yellow fever have killed more people than have all of man's wars. The bubonic plague, carried by a flea which lives on rats, swept through Europe during the Middle Ages, killing as much as one-third of the total population in a single epidemic.

However, it is a great mistake to consider all insects as pests. Most insects do not interfere with man, many are directly helpful, and all are part of the ecosystems that have evolved over the

41

millenia. For instance, we know that bees are absolutely essential to the life cycle of most flowering plants. In their search for food, bees inadvertently transfer pollen from flower to flower, thereby insuring fertilization. It is less generally known that many insects, such as species of springtails, are part of the process of decay which is necessary for recycling raw material. Insects are the prime food source of many animals that are vital, in turn, to the maintenance of natural balance. For example, the diet of many species of birds includes both insects and fruit. Fruit seeds transferred intact through the bird's digestive system and deposited at distant locations have been an important contributing factor in the continuing existence of certain plants. Thus, insects are necessary to the survival of many species of birds that are necessary in the life cycle of many fruit trees. In addition, many carnivorous or parasitic insects feed on insects that eat man's crops. (Recall that one need not eat "meat" to be a carnivore! An insect-eater is a carnivore.) The periodic invasion of some African villages by driver ants is a fascinating illustration of insect ecology. Many disease-carrying rodents and insects live in the village houses, posing a constant threat to the human population. At periodic intervals, however, millions of large driver ants invade the villages, chase away the inhabitants and eat everything that remains. When the people return, they find that their stored food supply is gone, but so are all the cockroaches, rats, and other pests—everything has been eaten. We do not maintain that driver ants are either good or bad; they are part of the natural order. Man, however, has found it advantageous to control certain pests which destroy large amounts of his crops.

It is important to emphasize at this point that "pest" insects have lived side by side with man for millions of years. Several factors have made this accommodation possible, if not pleasant. First of all, some insect populations have been controlled by natural predators. Also, many species of plants have evolved means of synthesizing various insecticides to protect themselves. For example, the balsam fir produces an insecticide that provides it with natural immunity to the linden bug; chrysanthemums make the insecticide **pyrethrum**; and the roots of some East Indian legumes contain the insect poison **rotenone**. But man now deems such natural accommodations insufficient. His population is so large and his food requirements are so pressing that he must compete to the death with insects. Physical barriers are usually insufficient for separating crops from insects; instead, modern agriculture uses poisons to protect crops. The remainder of this chapter discusses pesticides—their composition and use and their ecological consequences.

3.2　CHLORINATED HYDROCARBONS

In the late 1930's and early 1940's a new group of chemical pesticides were synthesized in various laboratories. These compounds do not occur in nature. Because they contain carbon, hydrogen, and chlorine they are called "chlorinated hydrocarbons." They now constitute a recognized class of chemical compounds. Some are commonly known by their trade names, such as DDT, aldrin, chlordane, dieldrin, endrin, and heptachlor. These pesticides were at first hailed as one of the great discoveries of the 20th century. They are easy and inexpensive to manufacture and are very effective poisons. DDT, for instance, can be used against any type of insect with some degree of success.

In the United States, the food loss attributable to insects is estimated to be about 10 per cent of the total crop, while the loss in many underdeveloped countries is estimated to be 30 to 40 per cent. (However, much of this is storage loss, which could be minimized by better physical isolation of grains after harvest.) DDT and similar compounds were therefore hailed as ways to rescue the world's food supply. In addition, mosquito control programs using DDT were seen as a means to reduce disease in areas of the world where over one billion people live. Before we point out some of the dangers of chlorinated pesticides, we must emphasize that the proper use of DDT has saved many millions of people from immediate death from disease, starvation, or both. Why, then, is there a concerted campaign to ban the use of DDT and related pesticides? There are three main arguments against chlorinated hydrocarbons:

(a) They are universal poisons.

(b) They degrade slowly.

(c) They are fat-soluble.

(a) Chlorinated hydrocarbons are universal poisons. The details of the action of chlorinated hydrocarbons are not completely understood, but it is known that they are non-selective animal poisons. These pesticides kill not only the insects, but also fish, birds, invertebrates, and mammals (including man). For example, in 1954, several communities in eastern Illinois were sprayed aerially in an effort to stop the westward movement of the Japanese beetle. The result was that many species of birds were almost completely annihilated in the sprayed area, ground squirrels were almost eradicated, 90 per cent of all farm cats died, some sheep were killed, and muskrats, rabbits, and pheasants were poisoned. These unwanted side effects might have been considered a necessary price to pay for pesticidal success, but the cost did not yield the desired benefit; the Japanese beetle population continued its westward advance.

One additional consequence of non-selective poisoning is that carnivorous as well as herbivorous insects are destroyed. Remember that populations are controlled in balanced natural systems. However, the use of non-selective sprays can lead to the destruction of the natural controls on relative population sizes. As an example, when DDT and two other chlorinated hydrocarbons were used extensively in pest control in a valley in Peru, the initial success gave way to a delayed disaster. In only four years, cotton production rose from 440 to 650 pounds per acre. However, one year later, the yield dropped precipitously to 350 pounds per acre, almost 100 pounds per acre less than before the insecticides were introduced. Studies indicated that the insecticide had destroyed predator insects and birds as well as insect pests. In fact, although the population of pest insects had decreased at first, the pest population soon recovered and flourished once again, this time quite resistant to the pesticides.

Insights into the mechanism of this apparent contradiction teach much about the relative advantages and dangers of pesticides. There is a biochemical basis to the dire environmental consequences of chlorinated hydrocarbons: Excess pesticide is not generally excreted by animals. Therefore, DDT and similar pesticides are usually more effective against predators than against the pests to be controlled. To help understand the reason for this, let us suppose that a field is sprayed and that pest insects feed on the poisoned leaves. Because excess DDT is not excreted by the insects, the concentration of this chemical in their bodies becomes greater than that in the plant leaves. Moreover, because death may be delayed for days or even weeks after poisoning, many poisoned, but living, insects will be eaten by other insects who are their natural enemies. Thus, the predators eat a diet that is more concentrated in DDT than that of the herbivores, the original pests. In turn, when birds eat the carnivorous insects, their meals have an even greater concentration of DDT than that of the insects they are eating. Thus DDT is concentrated as it moves up the food chain.

One study examined the successive concentrations of the pesticide **toxaphene** in various parts of a particular marsh. Both the mud on the bottom of the marsh and the bodies of small invertebrates contained 0.2 ppm of toxaphene.* This concentration was not lethal to the invertebrates. The fish that ate them had toxaphene in their muscles in quantities up to 8 ppm. The pesticide residue level in the fatty tissue of fish-eating birds reached as high as 650 ppm.

---

*ppm = parts per million. This means that, for every million pounds (or any other unit of weight) of pond sediment, there would be 2/10 of a pound of toxaphene. While this may not sound like much, many poisons are effective in extremely low concentrations.

Another problem with pesticides is that they tend to become less effective after some years of use. The pesticide composition is, of course, the same, but its use tends to result in the development of an immune pest population. It appears as though the pesticide has diminished in potency. This gradual deterioration of pesticidal action necessitates the use of larger and larger quantities of pesticide to maintain the same effect. To understand the reason for this phenomenon, we must consider the nature of genetic adaptation. The chemistry of plants and animals changes from time to time as a result of random accidental alterations – mutations – of their reproductive cells. A mutant has a good chance of survival if its particular mutation protects it from a hostile environment. This mechanism of random mutation has allowed insects to adapt to their environment for millions of years, and it is this process that protects insects from pesticides. In areas where spraying of DDT is heavy, species of insects whose body chemistry is immune to DDT have developed. This genetic immunity to chlorinated hydrocarbons is an extremely serious problem. In 1945 it was reported that about a dozen species had developed resistance to DDT; by 1960 the number had increased to 137 species. Of these, 65 were crop-destroying species, and many of the others were disease-carrying. It is important to remember that immune parents tend to pass their immunity on to succeeding generations, rendering the old pesticide ineffective. This effect was directly demonstrated in an experiment in which DDT-resistant bedbugs were placed on cloth impregnated with DDT. They thrived, mated, and the females layed eggs normally. The young, born on a coating of DDT, grew up and were healthy. Attempts to change pesticides have in many cases simply produced strains of insects that are resistant to more than one chemical.

Insect resistance to poisons is a serious problem in itself, but it can be compounded if the pest becomes resistant and various predators do not. If this happens, the pests have a new biological advantage and can thrive in greatly increased numbers. This is favored by three factors:

(1) Insect pests are generally smaller and reproduce more frequently than their predators. Also, it takes more time for large organisms to grow than small ones. More reproduction means more mutation, and consequently the mutation rate of the pests as well as the chances for occurrence of mutations that give immunity to pesticides will be greater. Thus the pest species develop immunity faster than the predator species.

(2) Predators eat a diet richer in pesticides than that of the original pests. Therefore, given an equal amount of resistance, more predators will succumb.

(3) There are always fewer predators than herbivores (including pests) in an ecosystem. But for any species to have a small population is to run the risk of extinction, because it is easier for a group containing only a few individuals to be wiped out by some disaster than it is for a group containing many. Therefore, there is a greater chance that the species of herbivores (the pests, which are more numerous) will survive than the species of predators.

We can now reconstruct what happened in Peru. Although the pesticides caused a drastic decrease in pest population size, resistant mutants soon displaced the susceptible individuals. By this time the situation had become worse than it was originally, because the natural predators did not achieve immunity so well as the pests, and therefore the pests were under less control than ever.

Another effect of broad-range pesticides is that they sometimes create new pest populations. Consider the story of the spider mite in the forests of the western United States. The spider mite feeds on the chlorophyll of leaves and evergreen needles. Because in a normal forest ecosystem predators and competition have kept the number of mites low, they have never been a serious problem. However, in 1956 the U.S. Forest Service started a campaign to kill another pest, the spruce budworm, by spraying with DDT. The result was that the budworms died, but so did such natural enemies of spider mites as ladybugs, gall midges, and various predator mites. The next year the forests were plagued with a spider mite invasion. Although the spruce budworm had been controlled, the new spider mite problem proved to be more disastrous.

The use of universal poisons as pesticides thus constitutes another case where man, by ignoring basic principles of natural balance, has compounded his problems. The response to such failure has often been to increase the number of sprayings or the

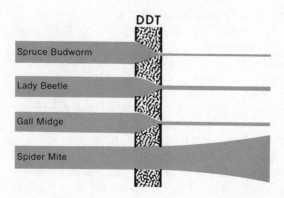

amount of insecticide applied per spraying. Obviously this "solution" may very well further compound the problem in time.

(b) Chlorinated hydrocarbons degrade slowly. Most chemical compounds which are naturally occurring are biodegradable—that is, they are degraded (broken down) by some form of life. Biodegradability is a phenomenon that has developed with the evolution of species; if there is energy or raw materials to be released by breaking down the molecule of a compound, then some organism has evolved to fit into the niche that can use that compound. However, many new compounds have been synthesized by chemists in recent years. Only some of these are similar enough to naturally occurring compounds to be biodegradable. Chlorinated hydrocarbons decompose slowly in nature, and many have a half-life of decomposition of 10 to 15 years.* The result is that these poisons exist long enough to produce many harmful effects.

We have already mentioned that chlorinated pesticides concentrate in predators high in the food chain. This characteristic is partly a result of the longevity of these chemicals, for it takes time to pass from insect to insect to songbird to falcon.

Because most fields that are sprayed at all are sprayed at least once a year, the concentration of DDT and similar compounds builds up in the soil. Soils of regularly sprayed farmlands have been found to hold as much as 15 to 20 pounds of DDT per acre. The danger of high concentrations of poisons in the soil arises from the fact that fertile soil contains much living matter. One pound of rich farm earth contains up to 1 trillion bacteria, 200 million fungi, 25 million algae, and 15 million protozoa, as well as worms, insects, and mites. These organisms are vital for continued fertility. They fix nitrogen, they break down rock and thus make minerals available to the plants, they retain moisture, they aerate the soil, and they bring about the essential process of decay. Without these organisms, the plants above ground would die. The effect on these organisms of an increasing concentration of poison in the soil is largely unknown. In many heavily sprayed areas of the world man is harvesting more food per acre than ever before. Yet some facts are coming to light which may presage future disaster. Studies in Florida have shown that some chlorinated pesticides seriously inhibit nitrification by soil bacteria. Termites have not been able to survive in soils that were sprayed with toxaphene ten years previously. As is the case for many types of ecological disruptions, the long-term results are not known. Perhaps bacteria are immune, or will become immune, to pesticide spraying. But the stakes in the

---

*The concept of half-life is discussed in Chapter 4.

gamble are large, for if the microorganisms in the soil die, large plants and animals cannot live.

(c) Chlorinated hydrocarbons are fat-soluble. When two substances are brought into contact with each other, they may mix so intimately that their individual molecules are dispersed in each other. Such a mixture is called a **solution,** and the two substances are said to be soluble in each other. For example, if 10 grams of salt are stirred with 100 grams of water, the resulting mixture is a true solution—it is so intimately mixed that it is equally salty throughout. Not all mixtures behave in this way. For instance, after oil and water are mixed two distinct layers can be observed, the oil on top, the water below. The two substances are said to be **insoluble** in each other. Substances that are mutually soluble have some chemical resemblance and some affinity for each other— therefore it is generally difficult to separate them. (It is easier to separate water from sand than from salt). Chlorinated hydrocarbons are soluble in fat but insoluble in water or blood. If small amounts of these compounds enter the body, they will concentrate in fatty tissue and cannot be washed away very efficiently by the blood. Thus, pesticide concentrations slowly build up in the bodies of animals.

The effect of this constant accumulation of poisons in animal tissue may range from undetectable to disastrous. The average American has a concentration of 12 ppm of DDT in his fat tissue. This level of DDT appears to have no detectable effect on his metabolism. However, long-term effects are unknown. Some data suggest that certain pesticides produce cancer. Larger concentrations which have been ingested by accident are directly fatal to man. However, the evidence seems undeniable that DDT poisoning is responsible for the sharp decline in populations of many birds, especially those that are secondary or tertiary consumers. One manifestation of the early stages of DDT poisoning is the inability to metabolize calcium properly. In birds, this has led to the production of thin-shelled eggs. Often these weakened eggs crack or break in the nest, with resulting prenatal death. Laboratory birds that are fed low concentrations of DDT in their diet all produce eggs with thin and weakened shells. The populations of several species of birds—among them, the peregrine falcon, the pelican, and some eagles—are declining so rapidly that many conservationists fear that they will become extinct in the near future. The effect of DDT on bone growth of human and other mammalian infants is still unknown.

Because of their fat-solubility, pesticides can be fatal to animals which store food energy in fat for use during winter months. Trout build up a layer of fat during the summer months when food is plentiful. In areas where the land has been sprayed, this fat contains high concentrations of DDT. During winter, the fat is used as a source of energy. The DDT released into the bloodstream upon fat breakdown has been known to kill the animal. The eggs of fish also contain a considerable amount of fat which is used as food by the unborn fish. In one case, 700,000 hatching salmon were poisoned by the DDT in their own eggs.

Because these pesticides are long-lasting and fat-soluble they can travel long distances. Tests on penguins living in Antarctica reveal traces of body DDT even though no one sprays crops anywhere on the continent. Undoubtedly the chemical has been carried there through a long succession of steps in the food chain.

We have just indicted pesticides on three counts: toxicity, stability, and fat-solubility. Why, then, should we use them at all? The primary, principal argument in favor of pesticides is that their use will increase man's food supply. Thus far, the total effect of chlorinated hydrocarbon pesticides on man's output of food has indeed been favorable because of the great increase in agricultural yields that has been achieved. However, ominous trends are developing. One, as already mentioned, is poisoning of the soil. The other is the ongoing poisoning of the earth's fisheries. Spraying of DDT in eastern Canada virtually annihilated a year's population of salmon in certain streams. Some fish were poisoned directly, because the streams in which they spawned contained high concentrations of the pesticide; others starved to death, because the insects on which they normally fed had been killed. Thus, part of a valuable source of man's food had been poisoned. In another case, pesticides dumped into the Mississippi River system by an industrial firm poisoned an estimated 10 to 15 million fish in a four-year period. In the Philippines, where milkfish farms produce a major source of protein for people of the region, aerial spraying against mosquitoes killed 60,000 fish in a single farm. Perhaps more ominous for the future is the effect of DDT on the phytoplankton in the sea. Phytoplankton are tiny plants which float freely in seas and are responsible for most of the primary oceanic food production. The sea without phytoplankton would be similar to the Earth's land mass without grains. It has been shown that DDT concentrations of a few parts per billion reduce the photosynthetic efficiency of these sea plants. Much higher concentrations have been reported in many coastal areas. Again, man is faced with a "trade-off." Is the risk to the seas worth the possible increase in land productivity?

An ideal solution, of course, would be to kill pests but not cause either direct or indirect harm to other species. The following sections describe several steps which have been taken in this direction.

3.3 OTHER METHODS OF PEST CONTROL

(a) Use of Short-lived Pesticides. Some classes of pesticides, such as the organic phosphates, are decomposed in nature on a time scale measured in days or weeks instead of in years. Such rapid disappearance offers great advantages, for these materials have less tendency to spread and move through the food chains. However, because they too are non-selective poisons, they also upset the ecosystem by destroying the natural enemies of the target pests. For this reason, the organic phosphates have caused many of the same harmful side effects as chlorinated hydrocarbons. Non-target animals (including man) have been killed, many insects have become resistant, and predator and parasite insect populations have been decimated.

(b) Use of Natural Insect Enemies: Predators, Bacteria, or Parasites. We have shown how destruction of predator populations by indiscriminate spraying has caused upsurges in pest populations. It is reasonable to assume that the opposite treatment—importation of predators—might be an effective control measure.

Excellent success with such techniques has been achieved in several cases. We mentioned earlier that DDT spraying against the Japanese beetle in Illinois caused widespread havoc among other species. The Japanese beetle, a native of the Orient, was inadvertently imported with a shipment of some Asiatic plants. The beetles thrived on the eastern seaboard of the United States and gradually became a major pest. Scientists then searched for natural predators and imported several likely species. One of these, an Oriental wasp, provides food for its young by paralyzing the Japanese beetle grub and attaching an egg to it. When the young wasp hatches, it eats the grub as its first food. The life cycle of the wasp is dependent upon the grub of the Japanese beetle; it does not naturally breed on the grubs of other insects. Therefore, this type of control is species-specific and does not seriously affect the rest of the ecosystem.

A similar technique has been in use since about 1885 for control of a certain scale insect which was virtually annihilating citrus groves in California. A parasite of the scale insect was found and was imported to the groves. This parasitic beetle brought the scale insect under control until the 1940's, when DDT was used to spray the orchards. The destruction of the beetle population by DDT led to a rapid, uncontrolled increase in the population of scale insects.

Other similar control programs include introduction of bacteria that destroy Japanese beetles (but almost nothing else) and importation of flies that parasitize the gypsy moth. There is, of course, a danger that an imported parasite might flourish and disrupt the ecosystem in some unforeseen manner.

(c) Sterilization Techniques. The screwworm (the parasitic larva of the screwworm fly) is a serious cattle pest and has been responsible for large financial losses to many ranches. Some years ago, the U.S. Department of Agriculture initiated a program to raise male screwworm flies, sterilize them by irradiation, and release them in their natural breeding grounds. The female mates of the irradiated males could lay no eggs. By the following spring the screwworm menace had been brought under control. This technique is another that is specific to the target pest and therefore not disruptive to the entire ecosystem. However, the method has had only limited success in controlling species other than the screwworm fly.

(d) The Use of Insect Hormones. Insects, at some point in their development, metamorphose from a larval to an adult stage. While an insect is in the larval stage it continuously produces a species-specific hormone. When the flow of that hormone stops, the animal metamorphoses. By spraying insects with the particular juvenile hormone of their own species, scientists have been able to inhibit metamorphoses. Because insects cannot survive or mate as larvae, such spray applications are eventually lethal. Moreover, because each species of insect is believed to have a unique juvenile hormone, a specific spray can, in theory, be developed for each species of insect. Research is currently under way to study in addition how hormones of insect pests may affect other insects that are their predators.

(e) Sex Attractants. In many species of insect the female emits a small amount of chemical sex attractant when she is ready to mate. The males detect (smell) very minute quantities of these chemicals and follow the odor to its source. Some success has been obtained by baiting traps with female sex attractants, thereby bringing the males to their death. Again, this type of control is highly specific.

(f) Use of Resistant Strains of Crops. It has already been mentioned that some plants are naturally resistant to pests because they synthesize their own insecticide. Plant breeders are now attempting to develop more such strains. A successful case has been the production of a variety of alfalfa which is resistant to the alfalfa weevil.

(g) **Cultural Controls.** Uniformity is not typical of virgin land masses. The systems created by modern agriculture differ from natural systems in that farms tend to specialize in a very few species of plants, whereas areas untouched by man do not. For instance, in Kansas and Nebraska, areas covering thousands of acres of land are covered almost exclusively with wheat fields. The result of such specialization is that a mold, a fungus, or an insect that consumes wheat has an almost unlimited food supply and an extremely hospitable environment. No barriers to spreading exist, and the pest can grow quickly in uncontrolled proportions. The advantage of diversity in nature was discussed in relation to the ability of an ecosystem to survive perturbations (see Chapter 1). One such perturbation is attack by specific consumers. A fungus that attacks wheat will spread more slowly if half the plants in a field are something other than wheat because the spores have less of a chance of landing on a wheat plant. Moreover, if the disease spreads slowly, there is more time for the development of natural enemies of the fungus or of strains of wheat that are naturally resistant to it. Therefore, one solution to the problem of pests is to grow plants in small fields, with different species grown in adjacent fields. However, this technique is less efficient than it is to plant, cultivate, and harvest large homogeneous fields.

Every method of pest control—and no pest control, as well—has both advantages and disadvantages. The hope for mankind is that knowledgeable and concerned people will make the correct decisions about how to feed the hungry adequately without upsetting natural balances and destroying the earth's future ability to produce.

PROBLEMS

1. What are the harms and the benefits that insects bring to man? By what mechanisms did man accommodate himself to insects before the production of modern pesticides?

2. What advantages are to be realized by the use of chlorinated hydrocarbon pesticides? What are the arguments against their use?

3. Explain why it often becomes necessary as time goes on to use larger quantities of a given pesticide to achieve the same results.

4. Pesticides have been known to be more harmful to predators than to the pests they are designed to control. What factors could account for this selectivity?

5. Compounds synthesized by man are less likely to be biodegradable than naturally-occurring compounds. Account for this difference.

6. What advantages and disadvantages would result from the use of a pesticide that was soluble in water? Do you think it likely that it would be practical to use a water-soluble pesticide? Defend your answer.

7. What methods of pest control are available as alternatives to the use of chlorinated hydrocarbons?

The following question requires arithmetic reasoning.

8. Data are given in Section 3.2 (a) on the increasing concentration of the pesticide toxaphene as it moves up the food chain from invertebrates to fish to birds. Are these increases constant, accelerating, or decelerating? Can you suggest a reason for this progression of concentration?

BIBLIOGRAPHY

The book that started much of our current concern about pesticides, and a more recent sequel to it, are the following:

Rachel Carson. *Silent Spring.* Boston: Houghton Mifflin Co., 1962. 368 pp.
Frank Graham, Jr. *Since Silent Spring.* Boston: Houghton Mifflin Co., 1970. 333 pp.

More detailed technical information is available from the following sources:

U.S. Department of Health, Education, and Welfare. *Report of the Secretary's Commission on Pesticides and Their Relationship to Environmental Health.* Washington, D.C., 1969. 677 pp.
Persistent Pesticides in the Environment. Cleveland: Chemical Rubber Company Press, 1970. 80 pp.

Legal and regulatory considerations are taken up in

Sandra C. Bloom and Stanley E. Degler. *Pesticides and Pollution.* Washington, D.C.: Bureau of National Affairs, Environmental Management Series, 1960. 99 pp.

4

RADIOACTIVE
WASTES

4.1   THE PROBLEM

Man has handled large quantities of radioactive material since
the discovery of atomic (nuclear) fission in 1939. Most notably, he
has developed, tested, and employed atomic bombs and other
atomic (nuclear) weapons, and he has constructed and operated
atomic (nuclear) reactors for the generation of power.

The production and disposal of radioactive wastes have necessarily been involved in all such activities. There are very sharp
differences in opinion as to whether or not the problem of radioactive wastes can be solved in an acceptable way, and as to whether
or not the benefits to man from his involvement with atomic materials will outweigh the disturbance to his global ecosystem. However, no one questions the fact that a waste problem exists. Furthermore, everyone agrees that it is impossible to invent anything to
prevent the production of these wastes from the various nuclear
processes that are now in use. It is important that we understand
how this situation differs from that of automobile exhaust, for
example. The pollutants (such as carbon monoxide) produced by
automobiles could be eliminated if suitable devices were invented.
We do not yet have an automobile that produces no pollution at all,
but it is theoretically possible to develop one that will discharge
carbon dioxide and water (the same products as those of respiration
of animals) but no carbon monoxide or other pollutants (see Chap-

54

ter 5). Some scientists think that such a task would be so difficult that it probably could not be accomplished in a reasonable time, but it is certainly theoretically possible. The production of carbon dioxide, CO_2, however, is a *necessary* consequence of the complete burning of gasoline in air; no one can invent a system or device to prevent this. The production of radioactive wastes as a result of our present atomic (nuclear) processes is inevitable in the same way that the production of CO_2 from the complete burning of carbon, $C + O_2 \rightarrow CO_2$, is inevitable. Since the production of waste cannot be prevented, it must necessarily be introduced into man's global ecosystem. The only influence we have in this matter is in determining how and where this waste material will be introduced so that it will produce the least possible disturbance to life on Earth.

In this chapter we wish to explain some of the fundamentals of the various processes from which radioactive wastes (sometimes nicknamed **radwastes**) are produced. We also wish to explain the concepts necessary in making judgments of benefits versus costs to man in relation to atomic (nuclear) materials. To understand these matters, however, we must first learn some of the fundamentals of nuclear chemistry, beginning with the structure and the mass of the atom.

4.2 FUNDAMENTALS OF ATOMIC STRUCTURE

Mass is a fundamental property of matter that is somewhat difficult to define. It is perhaps best to think of the mass of a sample as the quantity of matter in the sample. Thus, the mass of a cat is (approximately) 100 times the mass of a mouse. This means that there is 100 times as much matter in a cat as there is in a mouse. On Earth the weight of a body is determined by its mass; therefore, we can say that a cat weighs 100 times as much as a mouse. If we place a cat on one side of a balance, and 100 mice on the other, the balance will indicate that the weight (mass) of the cat equals the weight (mass) of 100 mice. However, this experiment must be done on Earth. In a space ship, there is no gravity and therefore no weight. But there is mass, and in a space ship the mass of a cat is still 100 times the mass of a mouse.

Atoms, too, have mass, although we cannot weigh atoms individually on a balance. Atomic masses may be expressed as whole numbers called **mass numbers.** The mass number of the ordinary hydrogen atom is arbitrarily taken to be 1. One kind of oxygen atom has 15.99 times as much mass as the hydrogen atom; the mass number of this oxygen atom is defined as the whole number nearest to 15.99, which is 16.

The Atomic Nucleus. Imagine that you had a piece of solid gold that was exactly the size and shape of this book (if the book were closed). This sample of gold would, of course, have a certain mass (weight). If you tried to stick a pin through the thickness of this gold book, you would find it difficult. The difficulty would be about the same wherever you tried to stick the pin. The conclusion of this crude experiment is that, so far as you can tell, the matter in a piece of solid gold is distributed equally in all the space occupied by the gold. Now imagine that you repeat the experiment in a much more delicate way. You hammer the gold into a very thin sheet (much thinner than a page of this book). You now try to puncture this thin gold sheet, not with a pin, but with a beam of very small particles* and you observe whether the particles go through the gold sheet or bounce off. The result of this experiment is that most of the particles go right through the gold sheet, but that a small proportion (one in several thousand) bounce off. The conclusion of this experiment is just the opposite of that of the crude test with the pin described earlier. The results of this new experiment tell us that matter is *not* distributed uniformly in a solid piece of gold, but is concentrated here and there in dense masses. The particles that bounce off the gold sheet must be those that happen to hit the dense masses.

This and similar experiments provide the fundamental evidence from which we conclude that the mass in any sample of matter is concentrated in what we call **atomic nuclei.** We observe the same kind of result with other types of thin sheets, such as those made of copper or of silver. We believe that the mass of all matter as we know it on earth is concentrated in atomic nuclei.

We have discussed atomic nuclei before discussing atoms, which is quite the opposite of what you will generally find in textbooks. But we are specifically concerned with nuclei in this chapter, and that is what we wish to emphasize. At this point, however, it is necessary to review some information about the atom itself.

We are familiar with the existence of electricity from such natural phenomena as lightning, sparks obtained by friction between different kinds of matter, and electric currents from living organisms, such as eels. We also know that there are two kinds of

---

*The particles originally used in this experiment were alpha particles (nuclei of helium atoms). The work was done by Lord Rutherford in 1911. It is not necessary to know the details of this experiment to understand this chapter.

electricity,* which we call positive and negative. From the results of experiments in which electric currents are passed through samples of matter** we also know that the electrical properties of matter are directly associated with the fundamental units that determine the chemical properties of matter. Evidence of the type obtained from electrochemical experiments, plus much evidence of many other kinds, supports the conclusion that all matter of the kind we are familiar with on Earth can be considered to be collections of fundamental units that we call **atoms**, and that these atoms consist of electrically negative, positive, and neutral particles. These particles, respectively, are electrons, protons, and neutrons.

PARTICLE	ELECTRICAL CHARGE	MASS NUMBER
Electron	−1	0
Proton	+1	1
Neutron	0	1

The nucleus of the atom contains the protons and the neutrons, and hence all of the positive charge and practically all of the mass. (The electrons do account for a small portion of the mass of the atom; the actual mass of an electron is about 1/2000 of that of the proton). The nucleus occupies comparatively little volume, however. If the nucleus of a gold atom were as large as the dot on the letter i, then the distance between two gold nuclei in a piece of solid gold would be about one foot. The electrons of the atom contain all the negative charge and are dispersed over most of the space, but they contain very little of the mass. An atom contains an equal number of protons and electrons; hence it is electrically neutral.

Atomic nuclei can be described by two numbers, which are defined as follows:

Atomic number = the nuclear charge
= the number of protons in the nucleus.

Mass number = the number of protons in the nucleus plus the number of neutrons in the nucleus.

A **chemical element** is a substance that consists of atoms of the same atomic number. Elements are considered to be kinds of ultimate chemical raw materials; all substances are composed of

---

*This information originally was derived largely from the experiments of Benjamin Franklin.
**These experiments were originally conducted by Michael Faraday in 1833.

elements. Each element has a common name and is represented by one or two letters. Eighty-eight elements occur naturally on Earth. These are elements 1 to 92 (with numbers 43, 61, 85, and 87 missing). The four missing elements, together with twelve more above 92, have been made by man, so that the total number now known is 104. Here are some examples:

NAME	NUCLEAR COMPOSITION		ATOMIC NUMBER (protons)	NUMBER OF ELECTRONS	MASS NUMBER (protons +neutrons)	SYMBOL
	Protons	Neutrons				
Carbon-12	6	6	6	6	12	^{12}C
Carbon-14	6	8	6	6	14	^{14}C
Calcium-40	20	20	20	20	40	^{40}Ca
Calcium-44	20	24	20	20	44	^{44}Ca
Uranium-235	92	143	92	92	235	^{235}U
Uranium-238	92	146	92	92	238	^{238}U
Uranium-239	92	147	92	92	239	^{239}U

Note that each element is symbolically represented by its letter(s) and its mass number.

Note also that a given element may have more than one mass number. Atoms of the same element (that is, atoms with the same atomic number) that have different mass numbers are called **isotopes**. Thus, ^{12}C and ^{14}C are carbon isotopes. Both ^{12}C and ^{14}C represent carbon atoms (or carbon nuclei), because they both have 6 nuclear protons. They are isotopes because they have different mass numbers. This difference results from the fact that there are different numbers of neutrons in the two nuclei. Isotopes of an element are chemically equivalent (or very nearly so): they have the same ability to combine. Thus, ^{12}C and ^{14}C are both chemically the same substance, carbon. One of the properties of carbon is its ability to burn in air. Therefore, we may write the following equations for burning the carbon isotopes:

$$^{12}C + O_2 \rightarrow {}^{12}CO_2$$

$$^{14}C + O_2 \rightarrow {}^{14}CO_2$$

Note that the two carbon nuclei, ^{12}C and ^{14}C, are not altered by the chemical reaction, and therefore persist in the carbon dioxide molecules produced by combination with oxygen.

4.3   RADIOACTIVITY

Some atomic nuclei are unstable; they are called **radioisotopes,** and the substances in which they exist are said to be **radioactive.** Some of these unstable nuclei occur naturally on Earth; others have been made by man. An example of a naturally-occurring radioisotope is radium-226, or ^{226}Ra. Natural radioactivity was discovered accidentally in 1896, when the Frenchman Henri Becquerel found that uranium minerals emit a radiation somewhat similar to X-rays. Later, it was learned that the energy of this emission was stored somehow in the atomic nuclei themselves. The emissions were found to consist of electrically positive particles, electrically negative particles, and electrically neutral rays (similar to X-rays) that we call gamma-rays. When a radioisotope decomposes, a new atom is left behind. In some cases this leftover atom is stable (nonradioactive); in others it is radioactive. For example, the radium-226 nucleus decomposes to produce a positive particle (an alphaparticle; see footnote on page 56) and a new element, radon-222, which is a radioactive gas:

$$^{226}\text{Ra} \rightarrow \text{alpha particle} + {}^{222}\text{Rn}$$

Radon-222 also decomposes to produce an alpha particle and still another radioisotope, polonium-218. The decompositions continue for seven more steps until a stable isotope, lead-206, or ^{206}Pb, is formed.

Now we ask two related questions: First, if ^{226}Ra nuclei are unstable, why are there any left on Earth? Second, if we had just one atom of ^{226}Ra, containing 1 nucleus, when would it decompose? The second question cannot be answered directly. Think of the ^{226}Ra nucleus as an energetic bundle of electrically charged matter; it may or may not break apart at any time. But we do know what the chances are that the ^{226}Ra nucleus will decompose in any given minute, or day, or year, or century. To understand this idea better, let us leave the Ra nucleus for a moment and consider a more familiar example. A bird lays an egg and leaves it behind, unattended, in the nest. At any time a predator, for example, a lynx, may come along and destroy the egg. How long will the egg last? Obviously we cannot answer the question; the particular egg we are considering may be lucky or unlucky. But if we understand the system completely we can tell what the chances of egg destruction are for any given period of time. Let us say that in any given day one egg has a 50–50 chance of surviving; thus the chance of survival equals the chance of destruction. Now imagine that we see 64 such individual eggs, each with a 50–50 chance of surviving any given day. If we return a day later, we expect to find half of them destroyed, and half intact. Therefore we expect to find 32 eggs. After another day the number will be halved again, and only 16 eggs will be expected to remain; the third day we would expect to see eight, then four, and so on. Because half of the eggs are expected to be destroyed in any given day, this time interval (in this case, a day) is called the **half-life**.

The half-life concept applies to radioisotopes. One nucleus of a ^{226}Ra atom has a 50–50 chance of surviving in any given interval of 1600 years. This means that the half-life of ^{226}Ra is 1600 years. Therefore, if one gram of ^{226}Ra were placed in a container in 1971, there would be only one-half gram left after 1600 years (in the year 3571), and only one-quarter gram after another 1600 years (in the year 5171), and so on. This process is called **radioactive decay.**

The concept of half-life does not imply that after 1600 quiet years half of the ^{226}Ra will suddenly decompose. Since there is a 50–50 chance that any one ^{226}Ra atom will decompose in 1600 years, there is also a 50–50 chance that one ^{226}Ra atom in 1600 atoms will decompose in 1 year — or that one out of 1600 × 365, which is one out of 584,000, will decompose in 1 day, or that about one out of 50,000,000,000 will decompose in 1 second. But a gram of ^{226}Ra contains about 2,650,000,000,000,000,000,000 atoms. This number is so large compared with the proportion of atoms expected to decompose every second that the decay process appears to be continuous, and any nearby Geiger counter will respond by clicking all the time.

The rate at which alpha-particles are emitted from a sample of ^{226}Ra depends on the quantity of ^{226}Ra. Since this quantity is constantly decreasing, the rate of emission, or radioactivity, of the sample of ^{226}Ra is also decreasing. Remember, however, that ^{226}Ra produces other radioisotopes when it decomposes. Any sample that has been decomposing for some finite time will contain some of the original ^{226}Ra plus some of each of its radioactive "daughters," as well as some of the stable final product, lead-206. These radioisotopes have different half-lives, ranging from fractions of a second to about 20 years. Therefore the total radioactivity produced by a sample of radium together with its radioactive waste products is more than that produced by the ^{226}Ra sample alone.

The radioactivity present on Earth before the 20th century was derived from radioisotopes that have survived over the history of the Earth. Therefore, these radioisotopes must have very long half-lives. The half-life of natural uranium-238, ^{238}U, for example, is

4,500,000,000 years. The radiations from such materials plus the effect of radiation that comes to the Earth from outer space is called the **background radiation**. It has been proven that radiation accelerates the process of genetic mutation. Life on Earth has always existed in the presence of a background radiation and, in fact, radiation-induced mutations have been a factor in the development of species.

In recent years man has vastly increased the quantity of radioactive materials on Earth, and therefore the total radioactivity on Earth has increased. As we mentioned at the beginning of this chapter, we cannot invent anything to stop this radioactivity. It slows down by radioactive decay at a rate determined by the half-lives of the radioisotopes involved. To understand this situation better, we must consider how man has produced more radioactive matter, how increased radioactivity affects life on Earth, and what problems are involved in the disposal of radioactive wastes.

4.4 HOW MAN HAS PRODUCED MORE RADIOACTIVE MATTER

Lord Rutherford and Frederick Soddy first proposed in 1902 that radioactive decay results in the change of atoms of one element to atoms of another element.* Seventeen years after that, in 1919, Rutherford produced transmutation artificially, by exposing nitrogen to alpha-particles to produce oxygen-17:

$$^{14}N + \text{alpha-particle} \rightarrow {}^{17}O + \text{proton}$$

The ^{17}O, however, is not radioactive.

Fifteen years later, in 1934, Irène and Frédéric Joliot-Curie, Mme. Curie's daughter and son-in-law, bombarded boron with alpha-particles and produced nitrogen-13, which is radioactive:

$$^{10}B + \text{alpha-particle} \rightarrow {}^{13}N + \text{neutron}$$

^{13}N was the first artificially produced radioisotope. This process therefore resulted in a man-made increase in radioactivity on Earth. It was the first production of atomic waste. However, the quantity

---

*"Soddy . . . turned to his colleague and blurted: 'Rutherford, this is transmutation!' Rutherford rejoined: 'For Mike's sake, Soddy, don't call it transmutation. They'll have our heads off as alchemists.' Rutherford and Soddy were careful to use the term 'transformation' rather than 'transmutation.'" (*Scientific American*, August, 1966, p. 91.)

of radioactivity produced by an experiment such as the one cited above has an inconsequential effect on the Earth, because only very small quantities of radioactive matter are involved. The alpha-particles used to bring about the transformation come from naturally radioactive sources such as radium, and consequently their availability is limited.

All this was drastically changed by the discovery of the **nuclear chain reaction**, which occurs in nuclear fission. For our purposes, it is important to understand first what a chain reaction is. Then we shall see why the chain reaction makes the production of radioactive wastes a real problem for life on Earth.

A chain is a series of links. Think of the process of making a chain; it involves the successive addition of links. ⊙⊙⊙⊙⊙⊙⊙ The process of adding links to a chain is called **chain lengthening.** If the end of a chain links onto the beginning, it forms a cycle, and the chain ends. ⊙ This is one form of **chain termination.** If more than one link is added to a given link, various arms of the chain develop; this is called **chain branching.** ⊙⊙⊙⊙⊙⊰

A series of steps in a process that occur one after the other, in sequence, each step being added to the preceding step like the links in a chain, is called a chain process or **chain reaction.** Chemical chain reactions can also undergo branching. An example of a branching chemical chain reaction is a forest fire. The heat from one tree may initiate the reaction (burning) of two or three trees, each of which, in turn, may ignite several others. If the lengthening of one chain proceeds at a given rate, the production of 10 branches means that 10 reactions are going on at the same time, so that the rate has increased tenfold. A chemical chain reaction that continues to branch can produce an explosion. The condition under which a chain reaction just continues at a steady rate, neither branching nor slowing down, is called the **critical condition.**

The production of the atomic (fission) bomb and of nuclear reactors depends on branching nuclear chain reactions. The process is initiated when a neutron strikes a ^{235}U nucleus and can proceed in any of several different ways. Two examples are shown here:

^{235}uranium + 1 neutron → ^{142}barium + ^{91}krypton + 3 neutrons

^{235}uranium + 1 neutron → ^{137}iodine + ^{97}yttrium + 2 neutrons

Note the following important points about these equations:

(a) The reaction is started by one neutron, but produces two or three neutrons. These neutrons can initiate two or three new reactions, which in turn produce more neutrons, and so forth. This is, therefore, a branching chain reaction. Thus, repetition of the first reaction could be written as follows:

$$3\ ^{235}U + 3 \text{ neutrons} \rightarrow 3\ ^{142}Ba + 3\ ^{91}Kr + 9 \text{ neutrons}$$

$$9\ ^{235}U + 9 \text{ neutrons} \rightarrow 9\ ^{142}Ba + 9\ ^{91}Kr + 27 \text{ neutrons}$$

and so on.

(b)   The ^{235}U nucleus is split in half (roughly) by these reactions. This is called **atomic** or **nuclear fission**. Fission releases considerable energy. If the branching chain reaction continues very rapidly, we have an atomic explosion. If the chain branching is carefully controlled, energy can be released slowly, and we have a nuclear reactor which can be used for power production.

(c)   The equations are balanced because the sums of the mass numbers remain constant:

$$235 + 1 = 142 + 91 + 3$$

and

$$235 + 1 = 137 + 97 + 2$$

(d)   Fission reactions produce radioactive wastes. ^{142}Ba, ^{91}Kr, ^{137}I, and ^{97}Y, the products shown in the equations above, are all radioactive. Furthermore, the reactions represented by these equations are only two out of many that occur in atomic fission. Many different radioisotopes are produced by atomic fission. Also, the fission products, as a group, are much more radioactive than the uranium that is the raw material. As we mentioned previously, the half-life of the naturally abundant uranium isotope, ^{238}U, is 4½ billion years.* Many American homes contain old (pre–World War II) orange-colored kitchen pottery prepared from a uranium oxide pigment. The radiation level from such materials is low. (However, it would be well to get them out of your kitchen. Donate them to the nearest university.) But the half-lives of the fission products are much shorter; some are measured in centuries, some in years, days, minutes, seconds, or fractions of a second. Therefore their radiation levels are much higher. These materials, produced in variety and

---

*^{238}U is not itself a fissionable isotope but is the raw material for the production of plutonium-239, which is fissionable. This conversion is explained later, in Section 4.7.

abundance by nuclear chain reactions, are the atomic wastes that concern us.

4.5   HOW INCREASED RADIOACTIVITY AFFECTS LIFE ON EARTH

A central point to remember is that isotopes of the same element have, for all practical purposes, the same chemistry. For example, iodine is used in the thyroid gland in man's neck. This iodine is part of the essential chemical growth regulator **thyroxin,** whose formula is $C_{15}H_{11}O_4NI_4$. Natural iodine, which comes from sources such as shellfish, is practically all ^{127}I, a stable (non-radioactive) isotope. Thus the thyroxin in your body is $C_{15}H_{11}O_4N^{127}I_4$. As we have seen, radioactive iodine is a nuclear fission product. One of its radioisotopes is ^{131}I. The thyroid gland looks at the chemistry of iodine, not the nuclear composition; it does not know the difference between ^{127}I and ^{131}I. Therefore, if the radioisotope is present in a man's food, it will find its way into his thyroxin, to produce $C_{15}H_{11}O_4N^{131}I_4$. Of course, the radioactivity of the ^{131}I is the same in the thyroxin molecule as it is elsewhere. It is important, therefore, to consider what may happen to a person with excess radioactivity in his body.

The effect of radioactivity on life depends on two classes of factors: (a) the kind of radioactivity present (the intensity and the types of rays produced) and (b) the chemistry of the radioisotopes, which influences how the radioisotopes travel and, especially, how they travel through food chains.

Let us look at the effect of radioactivity itself. Refer first to the units in Table 4.1. The high energy associated with radioactivity produces chemical changes. These changes include alterations of living cells, and the changes are practically always harmful.* Large doses can be fatal to man. The relationships are shown in Figure 4.1. As noted in Section 4.3, life on Earth has evolved in the presence of the natural background radiation level. This radiation can be just as harmful as that from artificial sources; however, its level is low, and life on Earth adjusts to the mutations it causes. The few organisms damaged by this radiation do not survive competition with the undamaged ones and hence die out. The even

---

*There is some chance that any given change is beneficial, but that chance is very small. The occurrence of a favorable change in cells due to radiation would be something like having a new set of Shakespeare's plays printed and finding that various accidental printing errors had improved the literary quality of the plays.

TABLE 4.1. UNITS RELATED TO RADIOACTIVITY

UNIT	ABBREVIATION	DEFINITION AND APPLICATION
Disintegration per second	dps	A rate of radioactivity in which one nucleus disintegrates every second. The natural background radiation for a human body is about 2 to 3 dps. This does not include "fallout" from man-made sources such as atomic bombs.
Curie	Ci	Another measure of radioactivity. One Ci = 37 billion dps.
Microcurie	μCi	A millionth of a curie, or 37,000 dps.
Roentgen	R	A measure of the intensity of X-rays or gamma-rays, in terms of the energy of such radiation absorbed by a body. (One R delivers 84 ergs* of energy to 1 gram of air.) The roentgen may be considered a measure of the radioactive dose received by a body. The dose from natural radioactivity for a human being is 5 R during the first 30 years of life. A single dental X-ray gives about 1 R, a full mouth X-ray series, about 15 R.
Rad		Another measure of radiation dosage, equivalent to the absorption of 100 ergs per gram of biological tissue.

*See Appendix for a discussion of units of energy.

rarer organisms that are improved by mutation, of course, survive preferentially; this selective process is evolution. But when radiation levels become too high, the regulatory mechanisms whereby the damaged organisms are weeded out are overwhelmed and cannot cope with the disruption. In other words, so many organisms are damaged that the entire species faces extinction. The damage

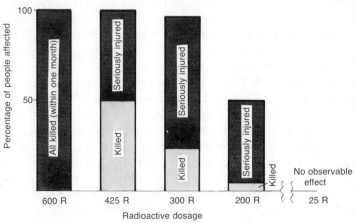

Figure 4.1. Effects of radioactivity on man.

done to individual beings, regardless of the effect on the species as a whole, must also be considered.

Radiation may affect any part of the human body. For example, radioactivity damages the blood by destruction of white blood cells and by injury to the bone marrow, the spleen, and the lymph nodes. Lung tumors, skin cancer, bone damage, sterility, and cataracts (clouding of the lens of the eye) are other specific effects that can be produced by large doses of radiation. The effects of small doses are more difficult to determine, but, according to our best present knowledge, even the smallest amount of intense radiation (one particle) can damage the nucleus of a single cell. A damaged cell, or one of its descendants, can become a cancerous one. If a germ cell is altered, the resulting genetic injury can be transmitted to future generations.

We have mentioned that the chemistry of the radioisotopes influences their mode of travel through the global ecosystem. Let us illustrate this point with reference to strontium-90, ^{90}Sr. This isotope is a radioactive fission waste product with a half-life of about 27 years; it is one of the components of "fallout," the mixture of radioactive wastes that are released from nuclear processes to the atmosphere and gradually settle to the surface of the Earth together with other miscellaneous dusts. Strontium (any strontium isotope, radioactive or not) is chemically similar to calcium, which is an important element in the bones of animals. In mammals, calcium is transmitted to the young through the mother's milk. Milk is thus an important source of calcium. Herbivores, such as cows, get their calcium from the vegetable matter of their diets. If a cow grazes in an area contaminated by fallout, the ^{90}Sr will be carried along with the calcium and will accumulate in the cow's bones and in her milk. This body process concentrates the ^{90}Sr, and there will be much more of it in a gram of milk than in a gram of grass. Cow's milk is used as a food by humans, especially by children, so children are likely to ingest relatively large quantities of ^{90}Sr. Lactating mothers concentrate ^{90}Sr in *their* milk, and as a result the nursing infant will ingest this radioisotope at his mother's breast. Concentration of radioisotopes in the food chain in the Arctic regions is particularly efficient. The effect of this concentration was noted when it was found that Eskimos absorbed more fallout radioactivity than did people who lived in other zones of the Earth where more fallout actually occurred. The first step in this highly efficient concentration is provided by the Arctic lichen, a plant that gets its mineral nourishment directly from dust particles that settle on it. For this reason, the lichen collects fallout dust particularly efficiently. In summer, caribou migrate north to the tundra where they wander over large areas in search of lichen, which becomes an important part of their diet. The effect is as if someone sent out the caribou to collect and bring back the ^{90}Sr, and they accomplish this task very well. Then, of course, the Eskimos eat the caribou, sometimes as their only food, and so, at the top of the food chain, they get the most concentrated radioactivity.

There are, of course, many other examples of radioisotope food chains in our global ecosystem. The point to remember is that the ecosystem, which is a biological mechanism, deals with matter according to its biochemical nature, not its isotopic composition. Therefore, it does not discriminate between radioactive and non-radioactive matter, and any radioactive element will always follow the same pathways as those of its stable isotopes.

A half-life of 27 years is short compared with one of 4½ billion years and is long compared with one of 1 second. What is important, however, is that 27 years is an ordinary (not very long or very short) length of time compared with the span of a human life. The radioactive dosage from a given quantity of ^{90}Sr will be reduced by only 50 per cent in 27 years. The dose from ^{238}U (4½-billion-year–half-life) will hardly be reduced at all, but it is not very much to begin with, so the total dose in a human lifetime is low. The dosage of a very highly radioactive material (such as is associated with a 1-second–half-life) is intense, but it doesn't last long and thus is not carried to a significant extent through a food chain. Therefore, a radioactive half-life that is in the range of the life span of living organisms can be particularly hazardous.

4.6   OPPOSING VIEWS ON NUCLEAR POWER

We have seen that radioactive wastes are necessary products of nuclear chain reactions, that radioactivity can damage living organisms, and that radioactive matter tends to distribute itself through an ecosystem as if it were not radioactive. Of course, man intervenes to control this distribution. He does not dump radioactive wastes into the environment in a random, accidental manner. He makes decisions and sets policies. These decisions and policies have become the subject of disagreements of the sharpest kind. At one extreme, the development and operation of nuclear power have been held to be a great boon to the future of mankind, providing inexpensive energy for a long time to come. The provisions for the disposal of the accompanying atomic wastes have been asserted to be so safe that the wastes pose no conceivable threat to man's welfare, certainly not in comparison to the benefits to be gained from the production of nuclear energy; even the proliferation of atomic weapons has been hailed as a guarantee of peace, because they effectively deter (nuclear) war. At the other extreme, these same developments have been called an imminent threat to man's survival as a species; the availability of cheap power from nuclear reactions has been said to be grossly exaggerated; the provisions for the safe disposal of atomic wastes have been denounced as insufficient and uncertain; the danger of radioactivity to man's health has been said to be consistently underestimated; and it has been recommended that all atomic weapons be scrapped and the entire nuclear reactor program abandoned at this time and that we fulfill all our energy needs from non-nuclear sources such as coal.

The questions involved in this debate are social as well as technical. To appreciate the problem, we must first understand in a general way how nuclear reactors work, what they do with their wastes, and what kinds of possible "trade-offs" this issue forces us to face.*

4.7   NUCLEAR REACTORS AND METHODS OF WASTE DISPOSAL

Let us start with the fundamental equations of nuclear fission:

$$^{235}U + 1 \text{ neutron} \rightarrow \text{fission products} + 2 \text{ to } 3 \text{ neutrons} + \text{energy}$$
(rare)   (1)

and

$$^{238}U + 1 \text{ neutron} \rightarrow \text{plutonium-239, or } ^{239}Pu$$
(abundant)   (2)

$$^{239}Pu + 1 \text{ neutron} \rightarrow \text{fission products}^{**} + 2 \text{ to } 3 \text{ neutrons} + \text{energy}$$
(3)

---

*We say "trade-off" in the sense of an exchange, or trade, between benefits and risks. The more benefits you get, the more it may cost.

**Recall that fission products, as a group, are highly radioactive.

Reactions (1) and (3) are the sources of power. Reaction (2) is not a source of energy, but utilizes the abundant but not fissionable ^{238}U to make a new fissionable element, ^{239}Pu. This element is the basis for the energy-producing reaction (3).

A nuclear reactor (Fig. 4.2) that produces power needs the following:

(a) A Means of Controlling the Neutrons. Too many neutrons bring about the danger of an uncontrolled branching chain reaction, which could create energy and waste products so rapidly that they could not be handled properly. Too few neutrons slow down the chain reaction and produce too little power. Control of the number of neutrons is effected by neutron-absorbing rods containing cobalt or boron that can be pushed into or pulled out of the reactor. Control of the energy of the neutrons, which influences the rate of the reactions they undergo, is exerted by a "**moderator**" – a substance that slows down neutrons without absorbing them. The moderator used in the first experimental reactor was graphite (a form of carbon); today's reactors use water.

(b) A Means of Removing the Heat Energy Produced. The energy yielded from the processes in the reactor must now be transferred to a circulating liquid, such as molten sodium metal. The heat of this liquid is then used to convert water to steam (in a heat exchanger, or condenser), which drives turbines to make electricity.

(c) A Means of Containing and Disposing of the Radwastes. The

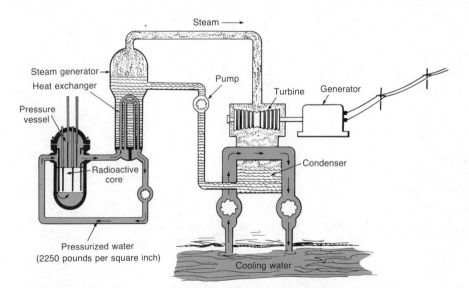

Figure 4.2. Schematic illustration of a nuclear power plant.

uranium, which is the "fuel," is inserted into the reactor in the form of long, thin cartridges. These fuel cartridges are coated (or "clad," as they say in the nuclear power plants) with stainless steel or other alloys. When enough impurities have accumulated, the absorption of neutrons by the fission products slows down the chain reaction and the fuel cartridges must be removed. It is here that the wastes are most concentrated, most radioactive, and therefore most dangerous. The spent fuel must be reprocessed, the uranium and plutonium recovered, and the wastes properly disposed of. The disposal method involves a complex series of operations. The ultimate objectives are to concentrate the radwastes as much as possible in order to save space and to store them where they will do no harm. But at first they are too hot to handle. (They are "hot" in two ways—they are highly radioactive, and they are boiling from their own released energy.) After some time, it begins to be possible to process the wastes; they must be concentrated and, when feasible, converted to a solid form that can be handled easily. This processing involves both physical and chemical stages and, of course, transportation from site to site. Figure 4.3 illustrates a decay curve for a hypothetical waste product whose half-life is one month. (Note that the radioactivity is decreased by half each month but that the curve is smooth; it does not go down in steps.) The kind of opera-

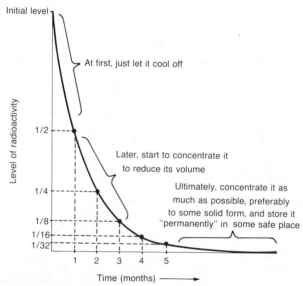

Figure 4.3. Disposal of radioactive wastes, for a hypothetical waste product with a one-month half-life.

tion that can be performed on the wastes therefore changes as time goes on.

The final step that is now considered to be best for ultimate disposal is to place the cooled, concentrated, solidified wastes in an abandoned salt mine or salt cavern that is expected to remain dry and undisturbed for thousands of years (Fig. 4.4).

A possible alternate procedure is to make up a kind of cement that incorporates the radioactive waste matter. This "hot cement" is then injected into underground seams of geologically stable rock,

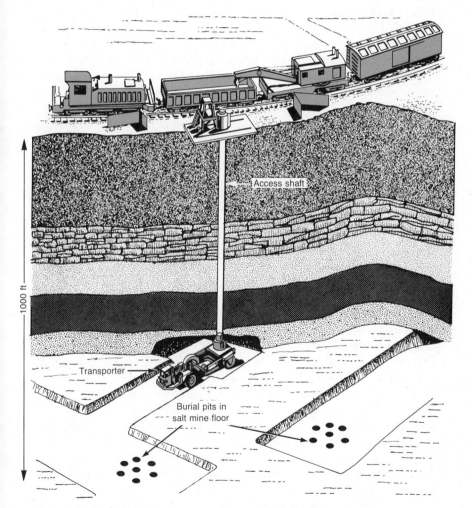

Figure 4.4. Final disposal of radioactive wastes in a salt mine. (© 1971 by The New York Times Company. Reprinted by permission.)

where it hardens and becomes part of the solid structure of the Earth's crust.

(d) A General Set of Safety Procedures to Insure Against Accidental Releases of Radioactive Material. The safety concept used by the nuclear power industry is that a triple layer of defense makes any serious accident almost unthinkable.

The first line of defense is the incorporation of safety factors in the basic design, construction, and operation of the reactor. For example, the nuclear fuel is not the pure metallic element uranium, but rather a ceramic form of uranium dioxide. This compound is much better able to retain most fission products, even when overheated. The fuel cladding is thus better protected against disruption. The uranium fuel itself consists of the natural non-fissionable isotope, ^{238}U, enriched with the fissionable ^{235}U by a factor only three or four times above its naturally-occurring level. This low level of enrichment provides an automatic protection against an increased fission rate if the temperature should rise accidentally, such as by failure of the coolant system. A modern nuclear reactor is therefore nothing like an atomic bomb, not even potentially so, even if all safety systems should fail. The use of ordinary water as the coolant and neutron moderator also provides automatic protection: an unexpected rise in power level would boil away some of the water, and this loss is used as a signal to initiate a reactor shutdown sequence. This description of the "first line of defense" is far from exhaustive; it is intended only to illustrate that safety must initially be inherent in the process itself.

The second line of defense is based on the assumption that the first line, contrary to all expectations, may somehow fail. It is rather like having a well-designed electrical system in a hospital to provide power for essential services such as operating rooms, and, in addition, having a standby gasoline-powered generator in case the primary system fails. Thus, if the mechanisms that cool the reactor core should fail, at least *two* other independent cooling systems are available. If the power system on which the emergency measures depend should fail, an off-site source of power can be used. If *that* fails, on-site diesel generators or gas turbines can take over. Secondary systems of this type are quite complex and are interrelated in such a way that their responses are specifically appropriate to the nature of the emergency. Furthermore, these responses are fully automatic; they do not have to be initiated by a human operator.

The third line of defense assumes that both the first and second lines may fail. This final barrier to radioactive contamination of the

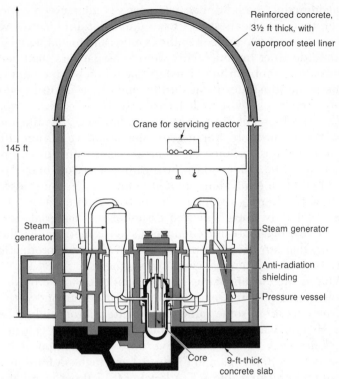

Figure 4.5. Containment structure for a nuclear power plant. (© 1971 by The New York Times Company. Reprinted by permission.)

environment is a massive containment structure (Fig. 4.5) which is a thick, vapor-proof, reinforced concrete housing that shields the reactor and steam generators. This barrier is designed to withstand earthquakes and hurricanes and to contain all matters that might be released inside, even if the biggest primary piping system in the reactor were to shatter instantaneously.

4.8   PROBLEMS AND ISSUES IN THE RELEASE AND DISPOSAL OF RADIOACTIVE WASTES

The questions at issue involve trade-offs at very high stakes for all mankind.

First, we will not consider the development and testing of nuclear weapons as a valid subject for cost-benefit analysis. If "mutual assured destruction" existing between two countries is to be considered a "benefit" because peace is guaranteed by fear of reprisal, then the cheapest way to accomplish this objective would be for each country to mine the other's cities and to maintain control over the "destruct button." This would cost far less than fleets of jets, missiles, or submarines. The fact that no such agreement would ever be made illustrates its absurdity, and if it is absurd to do this cheaply, it cannot be wise to do it expensively.

The development of nuclear power, however, is quite another matter. We can no more "renounce" power than we can agriculture. Therefore we must balance the nuclear risks against the air pollution that might be produced by the burning of gas, oil, or coal (fossil fuels) to yield an equivalent amount of energy. We must also consider how the Earth's resources are to be exploited. Gas, oil, and coal are not only fuels, but also essential raw materials for production of the many organic chemicals, such as medicinals, plastics, and dyes, that are useful to man. To burn them for production of energy is to waste them for chemical synthesis.

What *is* the current status of risks incidental to the production of nuclear energy? First, even with the best design and with accident-free operation, some radioactive matter is routinely released to the air and water outside the plant. This release can best be assayed in terms of the radioactive dose that it imposes on the average person. Refer again to Figure 4.1 and to the definition of the R (roentgen) unit in Table 4.1. A level of 0.17 R per year has been suggested as a "reasonable" maximum dose imposed by the nuclear power industry. How should we think of that level (which assumes that all goes well and there are no accidents) as a personal price to pay for the energy that we use? In six years, the exposure is equivalent to that of a single dental X-ray. Such a comparison may make the risk sound trivial, but we must recognize that such levels, whether from nuclear power plants or from medical sources, can account for a large number of deaths. This does not mean that the victim dies at the time of exposure, but rather that the death rate of exposed individuals is higher than that of non-exposed ones. But most Americans do not live near nuclear power plants and hence are not exposed to the maximum potential dose from such sources. It has been estimated that even an extensive American nuclear power industry would account for no more than about 1 to 5 deaths per year among people outside the plants. And remember, these deaths must be balanced against those that would be caused by air pollution if we used fossil instead of nuclear fuel.

Now, what about accidents? How safe are the safety devices? It is conceivable that a malfunction at one point could have far-reaching consequences, just as in a biological ecosystem. A loose piece of metal in the reactor could retard the flow of coolant, which could make the fuel elements overheat, which could rupture the cladding, which could interfere with the control rods, which could increase the flow of neutrons, which could further heat the fuel elements and melt them, which could release large quantities of radioactivity, which might or might not be adequately contained by the final barrier. Such series of events have been known to happen. A notable example was the fuel melt-down that occurred in 1966 at the Enrico Fermi Power Plant near Monroe, Michigan, not far from Detroit. This incident has been cited on both sides of the nuclear safety controversy. The opponents of nuclear power have emphasized the inherent uncertainty of even the most advanced safety procedures, and have pointed out the threat of major catastrophe to the city of Detroit. On the other side, the proponents of nuclear power have emphasized that the final lines of defense *did* work, because no radioactive cloud was in fact released.

Finally, how adequate are the methods of waste disposal? Before radwastes reach the final site, they do pose hazards in handling and transportation. Safety procedures outside the plant cannot generally be so rigorous as those inside. For example, massive containment barriers do not move along with trucks and trains. There are reputed, if not always well documented, instances of losses of radioactive matter in transit, either by carelessness, accident, or theft. Such material may include radioisotopes for medicine and industry as well as radwastes. It is difficult to assess the effects of such actual or potential losses; they are local, irregular, unpredictable, and their precise extent is unknown. Their control is largely a matter of improved enforcement of established procedures for safe handling.

The "permanent" storage poses questions of a different kind. Even if the repositories are safe for "thousands of years," we may ask whether such time spans are long enough. The waste products we store today may be relatively harmless after a few millenia, but if the waste disposal processes are continuous, there will always be some fresh material present, and so the danger does not decay. Man has lived on Earth for millions of years; are we setting a time limit to his tenure by a persistent interloping in the ecosystems of the Earth?

Or are we rather moving ahead toward ever more abundant sources of energy, particularly that which is potentially available from the nuclear fusion of hydrogen atoms so inexhaustively accessible in water, and which promises to be essentially non-polluting?*

---

*See Problem 11 for the equations for nuclear fusion.

PROBLEMS

1. Explain the difference between the mass and the mass number of an atom.

2. Define radioactivity; radioisotope; isotope.

3. What is a chain reaction? Explain chain propagation (lengthening); chain branching; chain termination; critical condition. Can the spread of a rumor among a large group of people function as a chain reaction? If so, illustrate how the chain could branch or terminate. Define the critical condition of such a system.

4. Outline the types of damage to the body that can result from exposure to high-energy radiation. Can there ever be any benefits? Explain.

5. What are the essential features of a nuclear fission reactor?

6. Sand-like radioactive leftovers from uranium ore processing mills, called "mill tailings," have been used to make cement for the construction of houses in Colorado, Arizona, New Mexico, Utah, Wyoming, Texas, South Dakota, and Washington. These tailings contain radium (half-life 1620 years) and its daughter radon (a gas, half-life 3.8 days), as well as radioactive forms of polonium, bismuth, and lead. Radon gas seeps through concrete, but is chemically inert. Are the following statements true or false? Defend your answer in each case:

   (a) Since radon has such a short half-life, the hazard will disappear quickly; old tailings, therefore, do not pose any health problems.

   (b) Even if the radon gas is present, it cannot be a health problem because it is inert and does not enter into any chemical reactions in the body.

   (c) Continuous ventilation that would blow the radon gas outdoors would decrease the health hazard inside such a house.

7. Outline the general concept and approach to safety used in nuclear power plants. Can you think of any specific series of events that would cause all of the safety features to fail and a radioactive cloud to be released to the atmosphere? If so, describe them.

8. Do you think it would be reasonable to set safety limits in nuclear power plants that would prohibit *any* release of radioactive matter? Defend your answer. If your answer is no, what criteria would you use to set the limits?

   The following questions involve some calculations.

9. Complete the following table by substituting the correct numerical value where a question mark appears. (Note that atomic numbers, but not mass numbers, appear in the table on page 210 of the appendix).

ISOTOPE	ATOMIC NUMBER	MASS NUMBER	NUMBER OF NEUTRONS IN NUCLEUS
Oxygen-18	?	?	?
Strontium-90	?	?	?
Uranium-?	?	?	141
Iodine-?	?	131	?

10. A Geiger counter registers 512 counts per second near a sample of radioactive substance. 48 hours later the rate is 256 cps (counts per second). What is the half-life of the radioactive substance? What rate will the counter register after an additional 144 hours?

11. The following equations represent the fusion of hydrogen nuclei to produce helium and to release a large amount of energy:

$$^2H + {}^xH \rightarrow {}^4He + 1 \text{ neutron}$$

$$^2H + {}^yH \rightarrow {}^3He + 1 \text{ neutron}$$

(a) What are the mass numbers of the hydrogen isotopes represented by x and y? What are the names of these isotopes?

(b) Can you suggest from these equations why nuclear fusion, when it can be controlled to produce useful energy, is expected to prevent a far lesser radwaste problem than that which results from fission reactors?

12. Cesium-137 is a radwaste whose half-life is 30 years. It is chemically similar to potassium, which is an essential element in plants and animals. Its compounds are readily soluble in water.

(a) How long will it take for 1000 mg of cesium-137 to decay to 125 mg?

(b) Since cesium compounds are soluble, would it be wise to dump this isotope into an open holding pond and let it dissolve and decay until only negligible quantities remain? Defend your answer.

ANSWERS

9. 8, 18, 10 neutrons; 38, 90, 52 neutrons; Uranium-233, 92, 233; Iodine-131, 53, 78 neutrons

10. 48 hours: 32 cps

11. x = 3, ^3H is tritium; y = 2, ^2H is deuterium

12. 90 years

BIBLIOGRAPHY

There are many books on atomic energy and nuclear engineering; many of them assume previous training in physics and chemistry. The following two texts, however, present somewhat more elementary introductions:

Alvin Glassner. *Introduction to Nuclear Science.* New York: Litton Educational Publisher, Van Nostrand-Reinhold Books, 1961.
Samuel Glasstone. *Sourcebook on Atomic Energy.* New York: Litton Educational Publisher, Van Nostrand-Reinhold Books, 1958.

Glassner's book is based on a short course given at Argonne National Laboratory since 1957, and presupposes only one year of college physics. Glasstone is more comprehensive and offers more introductory matter.
An excellent recent compilation of essays that deals with the various issues of radioactive wastes and presents both the reassuring and the alarming sets of viewpoints is

Harry Foreman, ed. *Nuclear Power and the Public.* Minneapolis: University of Minnesota Press, 1970. 272 pp.

The following books are directed almost exclusively to the negative aspects of nuclear power:

Richard Curtis and Elizabeth Hogan. *Perils of the Peaceful Atom.* New York: Doubleday and Co., 1969. 274 pp.
Sheldon Novick. *The Careless Atom.* Boston: Houghton Mifflin Co., 1968. 225 pp.

A brief discussion of methods of disposal of radioactive wastes may be found in

G. C. Collins. *Radioactive Wastes: Their Treatment and Disposal.* New York: John Wiley and Sons, 1961. 239 pp.

5

AIR POLLUTION

5.1   INTRODUCTION

Air is necessary for life on Earth. The addition of unwanted air-borne matter such as smoke changes the composition of the Earth's atmosphere, possibly harming life and altering materials. We call such atmospheric contamination **air pollution.** We usually reserve the word pollution for the alteration of the outdoor atmosphere by man's activities, although natural air pollution may result from non-human events such as pollen dispersal, volcanic eruptions, or forest fires ignited by lightning. We understand that air pollution can be controlled with filters and other technical devices. Why not, then, apply these control methods to prevent the release of contaminants into the atmosphere, and enjoy without penalty the human conveniences that technology brings us? Having read the first part of this book, you know that the problems produced by alteration of an ecosystem are subtle, complex, far-reaching, long-lasting, and sometimes delayed. When the ecosystem under consideration is the entire community of life on Earth, and when the disturbance affects so important a part of the ecosystem as the atmosphere, we know that the solution will not be simple.

We begin our discussion of air pollution with some fundamental concepts. First, we will review the arithmetic necessary to describe the composition of the air.

5.2 EXPRESSIONS OF CONCENTRATION; GASES AND PARTICLES

The concentration of a substance is the quantity of that substance in a given volume of space or in a given quantity of other matter. The concentrations of gases in air are usually expressed in ratios of volumes.* To illustrate, let us imagine that we make up a mixture that we will call "air" consisting of 20 liters of oxygen (O_2), and 80 liters of nitrogen (N_2). (This is not the same composition as the Earth's atmosphere, but close to it.) Then the composition will be 20/100, or 20 per cent, O_2 by volume and 80/100, or 80 per cent, N_2 by volume.

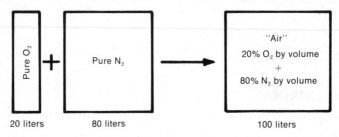

20 liters   80 liters   100 liters

(Assume constant pressure and temperature.)

Now imagine that we mix air with a pollutant, such as sulfur dioxide, SO_2. The concentrations may be expressed as follows:

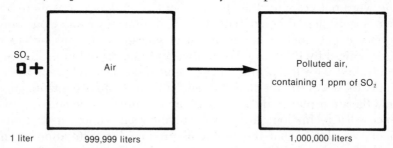

1 liter   999,999 liters   1,000,000 liters

In the illustration above, 1 ppm means 1 part per million, or 1 volume of SO_2 in 1 million volumes of polluted air. In dealing with even smaller concentrations, we sometimes use the expression "parts per billion," or ppb, which means volumes of pollutant per billion volumes of air.

For gases at constant temperature and pressure, the volumes occupied are directly proportional to the number of molecules con-

---

*Gases are substances that are dispersed in space as individual molecules.

tained. Therefore, to say that there is 1 ppm of SO_2 in the air means that one molecule in a million is a molecule of SO_2.

To understand the arithmetic is not enough. You should also get some feeling for the sizes of these numbers. Is 1 part per billion a large concentration or a small one? One billion pennies laid out rim to rim in a straight line would extend about 12,000 miles, or a distance almost equal to halfway around the Earth's equator. If one of these pennies were a bad one (a "pollutant") it would indeed be hard to find. This example makes 1 ppb sound small. But one cubic centimeter (this volume: ⬡) of air contains about 27,000,000,000,000,000,000 mole- cules. If this air were contaminated with 1 ppb of SO_2 (much less than in city atmospheres), there would be 27,000,000,000 (27 billion) molecules of SO_2 per cubic centimeter. This example makes 1 ppb sound large. Don't let either of these examples fool you. Concentration values by themselves do not provide information on the effects of pollutants. For some substances, 1 ppb is inconsequential; for others it is highly significant. We must therefore concern ourselves both with concentrations of pollutants and their effects on living organisms and on materials.

Not all air pollutants are gases. Some are airborne solid particles or liquid droplets, which are much larger bodies than individual molecules. For example, the diameter of a particle of dust may be 100,000 times that of a gas molecule. Concentrations of particles are usually expressed in terms of weight of pollutant per unit volume of air. Figure 5.1 illustrates the relative sizes involved.

We sometimes measure the rate at which particles settle out

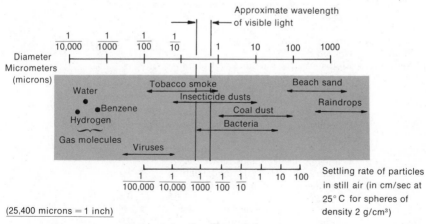

Figure 5.1. Small particles in air.

from the air. You may have read reports of air pollution expressed in units such as "tons of dustfall per square mile per month." However, such figures tell very little about total concentrations in air. Note from Figure 5.1, for example, that a 10-micrometer coal dust particle settles at about 1 centimeter per second, whereas a 1/10-micrometer smoke particle settles at a rate several thousand times slower, and gas molecules do not settle at all.

5.3 PURE AIR AND POLLUTED AIR

Now that we have the arithmetic in hand, let us look at the Earth's atmosphere and its pollutants. The most variable component of air is water vapor, or moisture, whose concentration may range from a negligibly small value in a desert to about 5% in a steaming jungle. If we neglect the moisture and consider only dry air, its composition by volume is roughly 78% nitrogen, 21% oxygen, and 1% of other gases. A more detailed breakdown is given in Table 5.1. Table 5.1 does not include non-gaseous, or "particulate," components. The "natural" concentrations of particulate matter in the air vary much more than those of gaseous matter.

TABLE 5.1. GASEOUS COMPOSITION OF NATURAL DRY AIR

GAS		CONCENTRATION (BY VOLUME)	
		ppm	per cent
	Nitrogen, N_2	780,900	78.09
	Oxygen, O_2	209,400	20.94
	Inert gases, mostly argon, (9300 ppm) with much smaller concentrations of neon (18 ppm), helium (5 ppm), krypton and xenon (1 ppm each)	9,325	0.93
"Pure"	Carbon dioxide, CO_2	315	0.03
Air	Methane, CH_4, a natural part of the carbon cycle of the biosphere; therefore, not a pollutant although sometimes confused with other hydrocarbons in estimating total pollution	1	
	Hydrogen, H_2	0.5	
Natural Pollutants	Oxides of nitrogen, mostly N_2O (0.5 ppm) and NO_2 (0.02 ppm), both produced by solar radiation and by lightning	0.52	
	Ozone, O_3, also produced by solar radiation and by lightning	0.02	

Thus, if we analyzed air in various parts of the Earth away from man's activities, the composition of the gases would be very close to the values in Table 5.1. But the particulate matter would vary widely from place to place. It would include nonviable (not capable of living) particles such as airborne soil granules, volcanic dust, and salts from evaporation of sea spray. It would also include viable particles such as plant and insect matter.

Some people define air pollutants to be substances not considered "natural" components of air. However, a hay fever sufferer may consider ragweed pollen an air pollutant even though it is a natural component of air in various parts of the Earth. We too would call pollen a pollutant, for we feel it is conceptually most satisfactory to define a substance called "pure air" and to think of any other component as a pollutant. Let us say that pure air is a gaseous mixture of the first six components of Table 5.1: nitrogen, oxygen, inert gases, carbon dioxide, methane, and hydrogen in the concentrations shown in the table, plus any additional moisture that may be present. Of course, any significant variation in these compositions could be harmful; for example, air containing 10% CO_2 would be poisonous, and air containing 10% H_2 or 10% CH_4 would be explosive. Thus, CO_2 in high concentrations is a pollutant. We will consider all other gases, regardless of concentration, whether of human or non-human origin, as well as all particulate matter, to be pollutants. This definition is, of course, arbitrary, but it will be convenient, and it is consistent with the thinking and practice of most people who are professionally involved with air pollution problems.

5.4 GASEOUS AIR POLLUTANTS

In this section we will describe the major classes of gaseous pollutants and some important individual pollutant compounds.

(a) Carbon Oxides. Carbon dioxide, CO_2, is a normal component of air (see Table 5.1) and a part of the carbon cycle of the biosphere; therefore, it is not ordinarily considered to be a pollutant. However, the burning of coal, oil, and natural gas as fuels produces large quantities of CO_2. The chemical equations are

$$\text{(burning of coal) } C + O_2 \rightarrow CO_2$$
$$\text{(burning of natural gas) } CH_4 + 2O_2 \rightarrow CO_2 + 2H_2O$$

It is estimated that the present rate of increase of the worldwide

CO_2 concentration is about 0.7 ppm per year. We must therefore consider the possible effects of a continued increase in CO_2 concentration on the Earth's atmosphere.

Carbon dioxide molecules, unlike the other components of pure air, have the property of absorbing the infrared (heat) radiation of the sun. Therefore, the more CO_2 in the atmosphere, the more heat the atmosphere can absorb. We do not know how severe the effect on the Earth might be. One of the most serious possible consequences would be a melting of the polar icecaps with consequent flooding of vast coastal areas throughout the globe.

Carbon monoxide, CO, is not a component of normal dry air, but is a product of the incomplete combustion of carbon or of carbon compounds.

$$2C + O_2 \rightarrow 2CO$$

The gas, though colorless, odorless, and non-irritating, is very toxic. The primary source of CO to which people are exposed in the outdoor atmosphere is automobile exhaust. The concentration level inside an automobile moving in a heavy stream of traffic on a multi-lane highway will be in the neighborhood of 25 to 50 ppm. The maximum allowable concentration for healthy workers in industry, for an eight-hour working day, is 50 ppm. A concentration of 1000 ppm can produce unconsciousness in 1 hour and death in 4 hours.

(b) *Compounds That Contain Carbon and Hydrogen, or Carbon, Hydrogen, and Oxygen.* The former category (carbon and hydrogen only) is the class of **hydrocarbons**. The latter group (carbon, hydrogen, and oxygen), sometimes called oxygenated hydrocarbons, or **oxygenates**, includes various classes, such as alcohols and organic acids. Such substances are introduced into the atmosphere by the incomplete combustion of carbon-containing fuels, along with the carbon monoxide mentioned above. Evaporation of liquids, as from the handling of gasoline or the spraying of paint, contributes to this pollution. The variety of effects from such substances is very great. Some of these materials are carcinogenic (cancer-inducing); some are irritating or malodorous; some undergo further chemical changes in the atmosphere to produce other pollutants; some are harmless.

(c) *Compounds That Contain Sulfur.* The important sulfur oxides are sulfur dioxide, SO_2, and sulfur trioxide, SO_3. From the viewpoint of its harmful effect on man and the difficulties involved in preventing its discharge into the atmosphere, SO_2 is probably the most significant single air pollutant. High SO_2 concentrations have been associated with major air pollution disasters of the type that have occurred in large cities, such as London, and that were responsible for numerous deaths. SO_2 is produced when sulfur or sulfur-containing fuels are burned:

$$S + O_2 \rightarrow SO_2$$

Since sulfur is present in coal and oil, the burning of these materials for heat and power produces SO_2.

The other important sulfur oxide, SO_3, is produced in the atmosphere by the oxidation of SO_2 under the influence of sunlight:

$$2SO_2 + O_2 \rightarrow 2SO_3$$

In addition, some SO_3 is introduced directly from combustion processes along with SO_2. The moisture in the air reacts rapidly with SO_3 to form a mist of sulfuric acid:

$$SO_3 + H_2O \rightarrow H_2SO_4$$

When such conversions occur, the material originally introduced to the atmosphere is called a **primary air pollutant**. The new materials produced by chemical reaction in the air are called **secondary air pollutants**.

Sulfuric acid is a very strong, corrosive acid that destroys living tissue, nylon stockings, and marble monuments. Sulfuric acid mist in air consists of droplets that are usually about 1 to 4 micrometers in diameter; this particular size range favors the deep penetration of the acid into the lungs, with consequent damaging effects.

Another important sulfur-containing compound is hydrogen sulfide, H_2S, which has the odor of rotten eggs. It blackens lead paints (Fig. 5.2), and is even more poisonous than carbon monoxide. H_2S is not a widespread pollutant like SO_2 or the hydrocarbons; its occurrence is usually associated with some specific source, such as decomposing organic matter, sewage, or some industrial operation. Some other compounds of sulfur are even more malodorous than H_2S.

(d) Compounds That Contain Nitrogen. The important oxides of nitrogen that occur in the air as pollutants are nitrogen oxide, NO, and nitrogen dioxide, NO_2. Both are produced by any combustion process that occurs in air, because some oxidation of atmospheric nitrogen occurs at flame temperatures:

$$N_2 + O_2 \rightarrow 2NO$$

$$2NO + O_2 \rightarrow 2NO_2$$

Figure 5.2. Paint damage from H_2S emitted from polluted San Francisco Bay waters. (Photo by J. E. Yocom. From Stern: *Air Pollution.* 2nd ed. New York: Academic Press, 1968.)

Thus, auto exhaust is a significant source of nitrogen oxides.

In considering the toxicity of these substances it is usually sufficient to focus one's attention on NO_2, because all other nitrogen oxides are converted to NO_2 in air. The effects of NO_2 on man range from unpleasant odor and mild irritation to serious lung congestion to death, depending on the concentration of the NO_2 and the duration of exposure. NO_2 concentrations in polluted outdoor air are not usually high enough to produce acute toxic effects, but they may produce or contribute to chronic effects, usually in the form of respiratory ailments.

NO_2 is also significant as an air pollutant because it is one of the key substances that enters into a chain of chemical reactions to produce "smog," discussed in the following section.

Some organic nitrogen compounds, called **amines**, are strongly malodorous, smelling like rotten fish.

(e) Ozone and Oxidants. Ozone, O_3, occurs to some extent in "normal" air (see Table 5.1), but in higher concentrations it is a toxic substance. The maximum allowable concentration for a healthy industrial worker over an eight-hour working day is 0.1 ppm.

It is somewhat curious that ozone has come to be popularly

associated with pure air and that devices which produce ozone have been regarded as "air purifiers." Ozone is, in fact, produced naturally in outdoor air by lightning, and its characteristic pungent odor under such conditions has probably been associated with the outdoors and with the cleansing action of rainstorms. These circumstances have led real estate developers to call their properties such fanciful but chemically inappropriate names as "Ozone Park," and the authors of travel folders to write such nonsense as ". . . A warm, a blue, an ever-changing sea, sparkling in the warm summer sunshine, and filling the air with ozone." (If it were true, the tourists would all be dead.) More serious and less laughable are the home and hospital appliances that purport to purify the air by producing ozone. Again, the confusion is rooted in some well-known but irrelevant chemical phenomena. Ozone is a chemically reactive substance that is sometimes used to deodorize exhaust gases (such as those emanating from sewage treatment) by oxidizing them to less objectionably odorous products before they are released to the atmosphere. The ozone concentrations needed for this treatment range between 10 and 20 ppm. Such concentrations would be rapidly fatal to human beings. The ozone produced by home appliance devices, on the other hand, is too dilute (around 0.1 ppm) to affect ordinary household odors to any measurable degree. Ozone has also been known for many years as a germicidal agent, and ozone generators are therefore sometimes used in the hope that they will suppress the spread of infectious organisms. The practical value of this scheme was questioned as early as 1913, when it was shown that ozone concentrations capable of killing bacterial pathogens killed guinea pigs even more rapidly. A more recent U.S. government study has concluded that "ozone, in low concentrations which do not otherwise cause irritation of the human respiratory tract, cannot be expected to provide any effective protection against airborne bacterial infection through direct inactivation of the infectious carrier particulates."[*] Thus, home appliances that produce ozone do not purify the air, but pollute it.

There are various other pollutant gases chemically related to ozone that are collectively called **oxidants**. The properties they have in common include certain toxic and irritating effects on people, various patterns of damage to vegetation, and the ability to produce cracks in natural rubber. (See Fig. 5.3.) These materials are generally produced by the reactions of hydrocarbons and other organic vapors with oxides of nitrogen in sunlight, and hence are typical pollutants of the atmospheres of sunny urban areas with

---

[*]U.S. Department of Health, Education, and Welfare, National Air Pollution Control Administration, *Air Quality Criteria for Photochemical Oxidants* (Washington, D. C., 1970), p. 6–18.

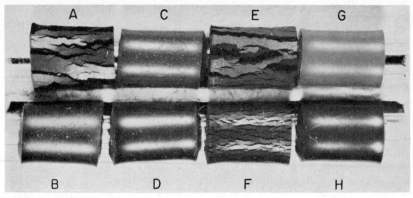

Figure 5.3. Effect of ozone exposure on samples of various rubber components. *A*, GR-S; *B*, Butyl; *C* and *D*, Neoprene; *E*, "Buna-N"; *F*, Natural rubber; *G*, Silicone; *H*, "Hypalon." (Photo courtesy of F. H. Winslow, Bell Telephone Laboratories. From Stern: *Air Pollution.* 2nd ed. New York: Academic Press, 1968.)

considerable automobile traffic, such as Los Angeles. The visible component of such pollution is commonly called "smog."

Chemical formulas of ozone and a typical chemical oxidant

Ozone, O_3

Peroxyacetyl nitrate, $C_2H_3O_5N$, an oxidant component of "smog"

(f) Hydrogen Fluoride, HF. This gas is an important pollutant because it has been shown to cause serious and widespread damage to vegetation. However, it is not a general component of polluted atmospheres, but rather originates from various specific industrial operations, such as the production of aluminum.

5.5 PARTICULATE AIR POLLUTION

The names used to describe airborne particulate matter are somewhat confused and inconsistent, referring sometimes to size, sometimes to source, and sometimes to a solid or liquid state. The following classification is a rough but useful description of this matter.

DIAMETER LESS THAN 1 MICROMETER*		DIAMETER GREATER THAN 1 MICROMETER*
Aerosols	May be solid or liquid,	Dusts (solid particles)
Smokes	depending on their	Mists (liquid droplets)
Fumes	origin	

*25,400 micrometers = 1 inch. A micrometer is also called a micron (see Appendix).

The word **aerosol** is used very commonly and should be remembered; it refers generally to any small particle in air. The 1-micrometer distinction between aerosol and dust or mist is not at all precise; many workers in air pollution refer, for example, to 10-micron (micrometer) aerosol particles.

Particle size is positively related to the speed of settling (see Fig. 5.1). Large dust particles are therefore troublesome only at relatively short distances from their source. Very small particles settle so slowly that they persist in the air for long periods of time and may be carried over long distances, often many miles.

Particulate pollutants may interfere with the transmission of heat from the sun to the Earth by reflecting a portion of the sun's rays away from the Earth. We do not know how extensive such heat loss might become if the particulate pollution of the atmosphere increases. A serious loss of the sun's energy would ultimately lower the average temperature of the Earth to produce another ice age. This effect is directly opposite to that of the heat absorption by atmospheric CO_2. It has been jokingly suggested that the loss of solar radiation by reflection from pollution might just balance the gain in heat from increased CO_2 concentration. This is indeed a joke; there is no prospect of man controlling such a delicate balance of cosmic processes.

There is great diversity among the types of particles in air. It is convenient for purposes of discussion to classify them into three categories: viable (capable of living), nonviable, and radioactive. We shall discuss the first two; radioactivity has been taken up in Chapter 4.

(a) **Viable Particles.** These include pollen grains, microorganisms such as bacteria, fungi, molds, or spores, and insects or parts of insects such as hairs, wings, and legs. Viable particles are responsible for many effects that are detrimental to man, including hay fever, some kinds of bronchial asthma, various fungous infections, and airborne bacterial diseases.

(b) Non-viable Particles. This group comprises a vast array of materials, some from natural sources and others from man's activities. The natural materials include sand and soil particles, salty droplets near the seashore, volcanic dust, and even particles of extraterrestrial origin. Man-made particulate pollutants include both organic and inorganic matter.* Much of the organic particulate matter is in smoke from the combustion of coal, oil, wood, and garbage. These particles consist mostly of carbon and include various carcinogenic (cancer-producing) compounds. Other airborne organic particles are insecticide dusts and some products released from food processing and chemical manufacturing. Inorganic particulate matter originates largely from metallurgical operations, non-metallic mineral production industries, inorganic chemical manufacturing, and the lead used in gasoline.

From an air pollution viewpoint, the most significant metallurgical operations are those involved in the production of iron and steel, copper, lead, zinc, and aluminum. The particulate matter discharged to the atmosphere in any given metallurgical process is not the pure metal itself but one or more of its compounds, some of which may be poisonous to living organisms. Non-metallic mineral products include cement, glass, ceramics, and asbestos. Operations in the manufacture of these products that are prone to produce airborne particles are blasting, drilling, crushing, grinding, mixing, and drying. Other inorganic particulate pollutants are specific to various chemical manufacturing operations, such as acid mists from production of nitric or sulfuric acids and phosphate rock dust from the manufacture of phosphate fertilizers.

Lead is used as an antiknock agent in gasoline in the form of an organic compound, tetraethyl lead or tetramethyl lead:

$$
\begin{array}{c}
\text{H} \\
| \\
\text{H—C—H} \\
\text{H} \qquad | \qquad \text{H} \\
| \qquad\quad | \qquad\quad | \\
\text{H—C——Pb——C—H} \quad \text{Tetramethyl lead} \\
| \qquad\quad | \qquad\quad | \\
\text{H} \qquad | \qquad \text{H} \\
\text{H—C—H} \\
| \\
\text{H}
\end{array}
$$

_____

*An "organic" substance is one whose molecules contain C-C or C-H bonds; it is not necessarily viable.

The total quantities produced in the United States in 1962 were 494 million and 18 million pounds, respectively, of tetraethyl lead and tetramethyl lead. These lead compounds are mixed with some simple chlorinated and brominated hydrocarbons before being added to the gasoline. What happens to all this lead? About 70 to 80 per cent of it is exhausted into the atmosphere in the form of small particles (ranging from a few hundredths up to several micrometers in size) of lead compounds, usually lead combined with chlorine and bromine, such as $PbClBr$. Of the remaining 20 to 30 per cent, about half is scavenged into the lubricating oil and half is retained in the engine and exhaust system. We know that lead compounds are poisonous, but we do not know precisely how damaging to the environment are the lead particles from automobile exhaust.

5.6   THE EFFECTS OF AIR POLLUTION

The idea that polluted air can be harmful to man dates back at least to the Middle Ages and to the concept of poisonous airs, or "miasmas." The Italian expression for bad air is *mala aria,* from which is derived the word "malaria," a disease once erroneously associated with the odors of swamps rather than with the germs carried by the mosquitoes that breed there. More direct evidence of bad effects from polluted air began to accumulate after the first use of coal, around the beginning of the 14th century. The dark smoke, the unpleasant odors, the blackening of buildings and monuments (see Fig. 5.4), all clearly resulted from the addition of unnatural and unwholesome substances to the atmosphere. However, only in the past few decades have we begun to realize the extent and complexity of air pollution effects and the imprecise nature of our knowledge about them. We may classify these effects into five divisions: (a) reduction of visibility and other atmospheric effects; (b) damage to vegetation; (c) direct effects on man; (d) injury to animals; and (e) deterioration of materials.

(a) **Effects on the Atmosphere.** The first noticeable effect of air pollution is that it becomes more difficult to see, as illustrated by Figures 5.5 and 5.6. At times in London and in some American cities the effect has been severe enough to curtail the flow of traffic. The reduction of visibility is caused by the scattering of light by small particles in the air; the most pronounced reduction is produced by particles between 0.5 and 1 micrometer in diameter.

Pollutants can also affect weather by mechanisms such as fog

Figure 5.4. Old post office building being cleaned in St. Louis, Missouri, 1963. (Photo by H. Neff Jenkins. From Stern: *Air Pollution*. 2nd ed. New York: Academic Press, 1968.)

Figure 5.5. Air pollution in the form of visibility-restricting particulate materials engulfs San Francisco, California, on June 2, 1960. (San Francisco News-Call Bulletin photo. From Stern: *Air Pollution.* 2nd ed. New York: Academic Press, 1968.)

formation and reduction of the amount of sunlight reaching the earth. These processes are very complex and our knowledge about them is incomplete. Some of these effects are briefly discussed at the end of this book in relation to the operation of the supersonic transport airplane.

(b) Damage to Vegetation. There has been widespread damage to trees, fruits, vegetables, and ornamental flowers by air pollution. The most dramatic early instances of such effects were seen in the total destruction of vegetation by sulfur dioxide in the areas surrounding smelters, where this gas is produced by the "roasting" of sulfide ores, as shown in the following equation:

$$2CuS + 3O_2 \rightarrow 2CuO + 2SO_2$$

We now know that there is a wide variety of patterns of plant damage by air pollutants. For example, all fluorides appear to act

'. . . And from the top there'll be a
spectacular view.'

Figure 5.6. The Gateway To The West
arch, in St. Louis, Missouri. Left: before
its completion, a cartoonist speculates.
Right: the finished product. (Cartoon by
Engelhardt, in the *St. Louis Post-Dispatch*,
from U. S. Public Health Service Publication No. 1561: *No Laughing Matter*. Photo
courtesy of the *St. Louis Post-Dispatch*.)

Figure 5.7. The Copper Basin at Copperhill, Tennessee. A luxuriant forest once covered this area until fumes from smelters killed all of the vegetation. (U. S. Forest Service photo. From Odum: *Fundamentals of Ecology.* 3rd ed. Philadelphia: W. B. Saunders Co., 1971.)

as cumulative poisons to plants, causing collapse of the leaf tissue. Photochemical (oxidant) smog bleaches and glazes spinach, lettuce, chard, alfalfa, tobacco, and other leafy plants. Ethylene, a hydrocarbon that occurs in automobile and diesel exhaust, makes carnation petals curl inward and ruins orchids by drying and discoloring their sepals. The total annual cost of plant damage in the United States has been estimated to be close to one billion dollars.

(c) Direct Effects on Man. Much attention has been focused on several air pollution disasters that have occurred during the past half-century. The most notable American episode began on the cloudy, dead-calm morning of Tuesday, October 26, 1948 in Donora, Pennsylvania. This city, located on a bend of the Monongahela River, housed a sulfuric acid plant, a zinc production plant, and a steel mill, in addition to other industries. Fog seemed to pile up rapidly during the first day, and the next, and the disagreeably pungent-sweet taste of sulfur dioxide became increasingly intense. (Sulfur dioxide is unusual among air pollutants in that it can be tasted at lower concentrations than it can be smelled.) Of the 14,000 people living in the valley, some 600 became ill. The symptoms of cough and irritation of eyes, nose, and throat started for most of them on the second day. There was no general warning of emergency, however, and the factories continued to operate without slowdown. By the afternoon of Thursday, the third day, the

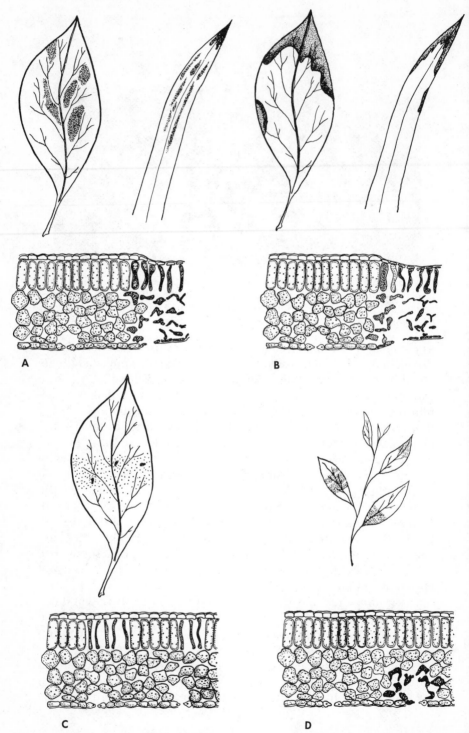

Figure 5.8. *See legend on opposite page.*

black fog had become so thick that it was difficult to see across the street, and all that was still visible of the mills was the tops of their chimneys, still discharging their pollutants into the air. Most of the 20 deaths that finally resulted from this episode occurred on the third day. This disastrous event, and the great London smog of December, 1952, which was responsible for 3000 to 4000 deaths, awakened many people to the dangers to health that air pollution can impose and prompted many studies of the problem. We understand now that such health effects can take many forms. These may be conveniently divided into three categories:

1. Acute illness, possibly leading to death.
2. Chronic disease, such as chronic bronchitis, pulmonary emphysema, or asthma. The precise relationships between such diseases and air pollution is often very difficult to establish; in most cases there may be more than one cause, such as the combination of air pollution and cigarette smoking.
3. General untoward symptoms and irritations, including overall discomfort, nervous impairment, eye irritation, and unpleasant reactions to offensive odors.

(d) **Injury to Animals.** The most serious effect in the United States has been the poisoning of livestock by fluorides and by arsenic. The fluoride effect, which has been the more serious, arises from the fallout of various fluorine compounds on forage. The ingestion of these pollutants by cattle causes an abnormal calcification of bones and teeth called **fluorosis,** resulting in loss of weight and lameness. Arsenic poisoning, which is less common, has been transmitted by contaminated gases near smelters.

(e) **Deterioration of Materials.** Acidic pollutants are responsible for many damaging effects, such as the corrosion of metals and the weakening or disintegration of textiles, paper, and marble. Hydrogen sulfide, H_2S, tarnishes silver and blackens leaded house paints. Ozone, as previously mentioned, produces cracks in rubber.

Figure 5.8. Effects of air pollution on plants. A, Sulfur dioxide injury. Note the blotchy interveinal areas on dicotyledon leaf and streaked areas on monocotyledon leaf (grass-type). B, Fluorine injury. Note the tip and edge necrosis on both types of leaves. Cross-section shows severe collapse and shrinking of internal structure. C, Ozone injury. Note the flecking or stippled effect on leaf. On sectioning, only the palisade layer of cells is affected. D, Smog-type injury. Note the change in position of the effect with the age of the leaf. On sectioning, initial collapse is in the region of a stoma. (From Stern: Air Pollution. New York: Academic Press, 1962.)

Particulate pollutants driven at high speeds by the wind cause destructive erosion of building surfaces. And the deposition of dirt on an office building, as on a piece of apparel, leads to the expense of cleaning and to the wear that results from the cleaning action. The total annual cost in the United States of these effects is very difficult to assess but has been estimated at several billion dollars.

5.7 THE CONTROL OF AIR POLLUTION

We now return to the question asked near the beginning of this chapter. If pollutants are being discharged to the atmosphere from a stack or an exhaust pipe, why not install some device that will remove the pollutants and allow only the harmless gases, such as air and moisture, to pass through? Such an objective is, in fact, entirely reasonable, and various types of air pollution control equipment are available. However, reduction of emissions at the point of discharge to the atmosphere is only one of several alternative approaches. In this section we shall briefly review some types of equipment that can be used for source control. The last section of this chapter will consider issues in the selection of processes that are less apt to generate pollutants.

There are two general classes of methods for controlling air pollution at the source: The pollutants are separated from the harmless gases and disposed of in some way other than by discharge into the atmosphere; or the pollutants are somehow converted to innocuous products that may then be released to the atmosphere.

CONTROL OF POLLUTANTS BY SEPARATION

Before we consider specific methods, we should realize that the separation of pollutants from a gas stream cannot be a final step in pollution abatement—the collected material does not disappear and therefore must be handled in some way. If the disposal of this residue is not considered, the solution of an air pollution problem may create a solid waste or a water pollution problem. However, such a conversion does at least ease the situation, because it is much more convenient to handle a small volume of solid matter than a large volume of air. (Typically, the volume of polluted matter in a contaminated air stream is in the range of 1/1000 to 1/10,000 of the total volume.)

Particulate matter may be retained on porous media (filters) which allow the gas to flow through. Such separations are possible because particles are much larger than gas molecules. For handling large gas streams, the filters are often in the form of cylindrical bags, somewhat like giant stockings, from which the collected particulate matter is periodically shaken out.

There are various mechanical collection devices that depend on the fact that particles are *heavier* than gas molecules. As a result, particles will settle faster and can be collected in a chamber that allows enough time for them to settle out. However, as evidenced by the data on settling rates shown in Figure 5.1, such methods are practicable only for very large particles. More important than their settling rate is the fact that heavier particles have more *inertia*. As a result, if a gas stream that contains particulate pollutants is whirled around in a vortex, the particles may be spun out to locations from which they may be conveniently removed. A device of this sort, called a **cyclone**, is shown in Figure 5.10.

Particles may also be electrically charged, and a collecting surface that bears an electric charge of the opposite sign therefore attracts them. Devices of this sort, called **electrostatic precipitators,**

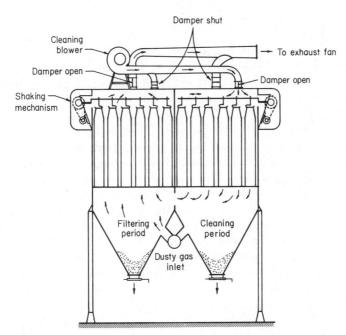

Figure 5.9. Typical bag filter employing reverse flow and mechanical shaking for cleaning. (From Stern: *Air Pollution.* New York: Academic Press, 1968.)

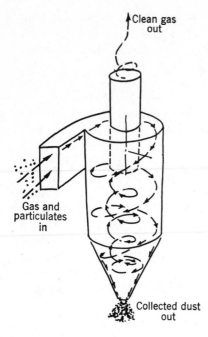

Clean gas
out

Gas and
particulates
in

Collected dust
out

Figure 5.10. Basic cyclone collector. (From Walker: *Operating Principles of Air Pollution Control Equipment.* Bound Brook, N. J.: Research-Cottrell, Inc., 1968.)

are used on a very large scale, notably for reducing smoke from power plants that burn fossil fuels.

Pollutant gases cannot feasibly be collected by mechanical means, because their molecules are not sufficiently larger or heavier than those of air. However, some pollutant gases may be more soluble in a particular liquid (usually water) than air is; they may therefore be collected by a process that brings them into intimate contact with the liquid. Devices that effect such separation are called **scrubbers.** The methods of making contact between gas and liquid include spraying the liquid into the gas and bubbling the gas through the liquid. Ammonia, NH_3, is an example of a gas that is soluble in water and can be scrubbed out of an air stream.

Gas molecules adhere to solid surfaces. Even an apparently clean surface, such as that of a bright piece of silver, is covered with a layer of molecules of any gas with which it is in contact. The gas is said to be **adsorbed** on the solid. "Adsorbed" means "held on the surface of a substance," and is different from "absorbed," which means "held in the interior of a substance." The quantity of gas that can be adsorbed on an ordinary piece of non-porous solid matter, such as a coin, is too small to be of any consequence as a means of collecting pollutants. However, if a solid is perforated

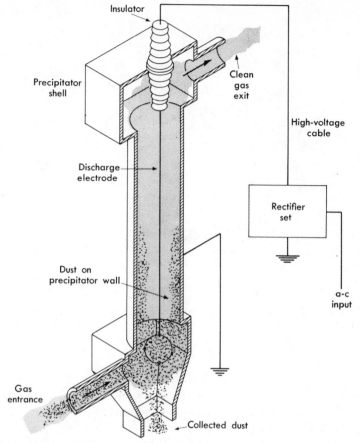

Figure 5.11. Basic elements of an electrostatic precipitator. (From White: "Industrial Electrostatic Precipitation." Reading, Mass.: Addison-Wesley.)

Figure 5.12. Schematic drawing of a spray collector, or scrubber. (From Stern: *Air Pollution.* 2nd ed. New York: Academic Press, 1968.)

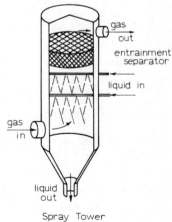

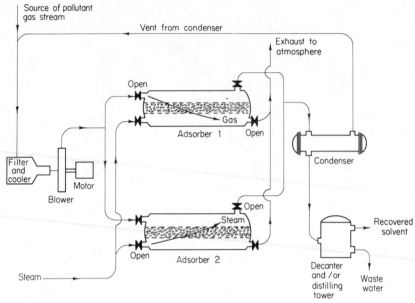

Figure 5.13. Two-stage system for adsorbing and collecting air-polluting gases.

with a network of fine pores, its total surface area (which includes the inner surfaces of the pores) may be increased so much that its capacity for gas collection becomes significant. Such a solid is **activated carbon**, which can have several hundred thousand square feet of surface area per pound. Activated carbon is made from natural carbon-containing sources, preferably hard ones, such as coconut shells or peach pits, by charring them and then causing them to react with steam at very high temperatures. The resulting material can retain about 10 per cent of its weight of adsorbed matter in many air purification applications. Furthermore, the adsorbed matter can be recovered from the carbon and, if it is valuable, recycled into the process or product from which it had escaped.

CONTROL OF POLLUTANTS BY CONVERSION

By far the most important conversion of pollutants is by oxidation in air. Oxidation is applied most often to pollutant organic gases and vapors, rarely to particulate matter. When organic substances containing only carbon, hydrogen, and oxygen are completely oxidized, the sole products are carbon dioxide and water, both innocuous. However, the process is often very expensive because considerable energy must be used to keep the entire gas stream hot enough (about 700°C) for complete oxidation to occur. If the pollutant is sufficiently concentrated, its own fuel value may contribute a large part of this energy. Also, the required combustion temperature may be reduced by using a catalyst.

There are a number of possible chemical conversions of pollutants other than combustion in air. These include the chemical neutralization of an acid or a base and the oxidation of pollutants by agents other than air.

5.8 PROBLEMS AND ISSUES IN THE CONTROL OF AIR POLLUTION

The problem of reducing air pollution goes beyond the question of what control method to apply at the source. The problem involves social and economic as well as technical factors. We will illustrate the nature of these issues by two examples.

The first example is the control of SO_2 air pollution from fuel combustion sources. We will not consider such other sources of SO_2 as copper smelters, petroleum refineries, sulfuric acid plants, steel mills, and volcanos.

We burn sulfur-containing coal, oil, or natural gas to produce heat and power. The sulfur in the fuel oxidizes to SO_2 and pollutes the air. The following control methods are conceivable.

Change to another energy source. Use hydroelectric energy, nuclear energy, or solar energy. Hydroelectric sources, however, are insufficient for present consumption. Solar energy is abundant, but not yet developed. As a partial solution, we could change to fuels with lower sulfur content, such as from a high-sulfur coal to a low-sulfur oil or gas. However, the supply of such fuels is limited.

Take the sulfur out of the fuel before it is burned. This approach is being investigated, but it would necessarily increase the cost of the fuel. For example, the reduction of the 5% sulfur content of crude oil to 0.5% was estimated in 1966 to cost about 30 cents per barrel. This price, which the consumer would ultimately pay, represents about 12% of the cost of a barrel of untreated oil.

Take the sulfur out of the exhaust gases before they are discharged to the atmosphere. A number of processes are being tried. It is difficult to predict their effect on power costs because this depends to some extent on the value of the recovered sulfur. In general, the systems require elaborate engineering and would significantly increase the investment cost of the power plant.

Use more efficient methods of fuel combustion. Any increase in efficiency of power production means that less fuel is used and hence less SO_2 produced.

Get along with less fuel per person. Instead of setting our thermostats at 74° and wearing light clothes indoors, we could set them at 64° and wear sweaters.

Decrease the rate of population growth. Fewer people would use less fuel, so less SO_2 would be produced.

This brief outline points out possible lines of approach to the control of a single air pollutant, SO_2, from a single type of source. The problems of estimating costs and balancing various trade-offs necessary to achieve a given reduction in pollutant concentration are difficult in themselves. What we are not sure about is how much reduction is needed to prevent specific adverse effects. In the air pollution disasters of Donora, Pennsylvania in 1948 and London, England in 1952, for example, the SO_2 concentrations, though unusually high, were not high enough to account in themselves for the reported deaths. It is thought that the combination of SO_2 and aerosols is more toxic than either pollutant alone. Recent work on the reactions of SO_2 in the air under the influence of sunlight indicate that other sulfur-containing gases are produced; some of these may be more toxic than SO_2 itself. Research on the health effects of air pollution has repeatedly shown that previously undetected contaminants may be toxic in surprisingly small concentrations. The criteria for acceptable air quality are therefore not fixed, but are subject to change as new findings emerge. Thus the problem of adequate control is complicated by incomplete knowledge of objectives and by various trade-off questions that must be answered if we are to relate the effectiveness of controls to their costs.

Our second example of this problem of relative values will emphasize the trade-off aspects of the problem. We will select that well-known source of air pollution, the gasoline-driven automobile, and an appealing solution, the electric car. The electric car is powered by batteries and is therefore practically non-polluting. However, the switchover would create various problems, such as the following:

(a) Need for More Central Power Plants. The energy for the electric cars isn't free. It would necessarily result in increased output by central power plants, which would be the source of the power required for recharging batteries. It is estimated that an increase of about 50%, costing over 20 billion dollars, would be necessary. Power plants are experiencing difficulty now in keeping up with the demand, especially during the summer (air conditioning) season. The required increase in capacity would take some time to establish, so the changeover would have to be gradual. Furthermore, power plants create their own pollution. This fact would be strong persuasion in favor of nuclear power plants, whose problems are discussed in Chapter 4. In general, however, it is expected to be easier to control emissions from a small number of large power plants than from millions of automobiles. Therefore the overall result should be a decrease in air pollution.

(b) Economic Depression in the Oil Industry. It is difficult to predict accurately how much loss would be suffered by the oil industry; this would depend in part on how many of the new central power plants used oil. A fair guess is that total sales would be cut in half. The result would necessarily be pockets of unemployment and depression, and the shift to new industries would involve dislocations.

(c) Changes in Automobiles and the Automobile Industry. It is not feasible to manufacture electric cars that are powerful enough to resemble the heavy, luxury-type gasoline-driven automobile. It is therefore likely they would resemble the new "mini" sub-compacts. There would be fewer "extras" sold because there would be less room for accessories in the smaller cars. This loss of profit for the manufacturer, added to the fact that battery components are expensive, might make the small electric cars cost about as much as or more than the larger gasoline-driven ones. The entire "parts" industry would have to change. Instead of furnishing spark plugs, mufflers, carburetors, and so forth, it would have to supply motor commutators, electrical controls, and the like. During the changeover, which would take years, both kinds of parts would be needed. In fact, the changeover might require the development of a hybrid vehicle — one that used both a battery-driven motor and a gasoline engine. The gasoline engine in such a vehicle would be small and would operate continuously, while the battery would provide peaks of power for acceleration and would absorb braking power from the motor. It is estimated that such a system would reduce noxious emissions by at least 50% compared with an all-gasoline vehicle of the same power. The use of intermediate designs of this type would greatly prolong the ultimate transition to an all-electric car.

(d) Changes in Service Stations. Mechanical repairs would give way to electrical repairs. Gasoline sales would have to be replaced by battery sale, rental, exchange, and recharging services.

(e) Changes in Highway Design. The current high-speed interstate highway system is designed for the rapidly accelerating gasoline-driven automobiles. The requirement for fast pickup is especially critical when an automobile enters the stream of swiftly moving highway traffic. The existing acceleration lanes would be insufficient for safe entry of the slower electric cars. This difference might prove to be especially difficult during the transition period when both types of vehicles were on the road.

(f) Taxation. The Federal government and the states gain revenue from gasoline taxes. The money from these taxes is generally earmarked for the building and maintenance of roads; thus the people who pay for the highway system are those who use it. With the use of electric cars and the consequent decrease in gasoline purchases, this income would have to be replaced, most probably by an increase in tax on electricity. This increase, however, would be imposed on everyone, and the non-driver would be paying money which would primarily benefit the operators of automobiles. Moreover, the resulting increased cost of electricity would influence people to avoid its use where possible. This would result in the use of fewer electric stoves and electric hot water heaters, for example. Instead, gas and oil would be used, leading to the same kind of pollution that the electric car was designed to eliminate.

This example should remind us that, like ecosystems, human civilizations also are delicately balanced systems—the products of cultural evolution. We must recognize that cultural disruptions, even for the sake of improving such a vital component of the Earth's environment as its atmosphere, have complex and serious social consequences.

PROBLEMS

1. What is air pollution? List ten gaseous and ten particulate air pollutants.

2. A concentration of 1 ppb of ethylene, (C_2H_4) gas in the air cannot be smelled and has no demonstrable effect on people, animals, or materials. However, it does produce dried sepal injury in growing orchids—to such a degree that they become unfit for sale. Is 1 ppb of C_2H_4 a high concentration, or a low one? Would your answer depend on whether or not you were an orchid grower?

3. Distinguish between primary and secondary air pollutants. Give three examples of secondary air pollutants.

4. Suppose that you keep some animals in a cage in your room and you are disturbed by their odor. Comment on each of the following possible remedies, or some combination of them, for controlling the odor:

(a) Spray a disinfectant into the air to kill germs;

(b) Install a device that recirculates the room air through a bed of activated carbon;

(c) Clean the cage every day;

(d) Install an exhaust fan in the window to blow the bad air out;

(e) Install a window air conditioning unit that recirculates and cools the room air;

(f) Install an ozone-producing device;

(g) Spray a pleasant scent into the room to make it smell better;

(h) Light a gas burner in the room to incinerate the odors;

(i) Keep an open tub of water in the room so that the odors will dissolve in the water

5. Define aerosol; dust; mist.

6. Differentiate between "oxygenate" and "oxidant." What is a hydrocarbon?

7. Differentiate among the terms organic, viable, and nonviable.

8. Of the five categories of air pollution effects discussed in Section 5.6, how many do you think could be produced by sulfur dioxide, SO_2, or by secondary air pollutants derived from SO_2? Justify your answer for each category.

9. Distinguish between separation methods and conversion methods for source control of air pollution. What is the general principle of each type of method?

10. Explain the air pollution control action of a cyclone; a settling chamber; a scrubber; activated carbon; an electrostatic precipitator; an incinerator.

11. Discuss the trade-off problems in the conversion from leaded to non-leaded gasoline. Include economic factors and problems that may arise during the changeover period. You may also take the following facts into account if you think they are relevant:

(a) If the non-leaded gas is to be equivalent in octane rating (ability to avoid engine knocking) to the leaded gas, the gasoline must be modified in some way or some non-lead antiknock agent must be found. Consider the possible consequences of these requirements.

(b) If the non-leaded gas is to be lower in octane rating, the automobiles must be modified in some way. This has already occurred for 1971 cars in the United States, whose engines are built for lower compression.

The following questions require arithmetic computation or reasoning.

12. Is the speed of settling of particles in air directly proportional to their diameters? (If the diameter is multipled by 10, is the settling speed 10 times faster?) Justify your answer with data from Figure 5.1. Is a settling chamber a good general method of air pollution control? Explain.

13. One liter of carbon monoxide, CO, is mixed with 99,999 liters of air, without changing the temperature or pressure. What is the concentration of CO in the mixture in ppm by volume? in ppb by volume?

ANSWER

13. 10 ppm; 10,000 ppb

BIBLIOGRAPHY

The basic text on air pollution is a three-volume work:

Arthur C. Stern. *Air Pollution.* New York: Academic Press, 1968. Volume I, 694 pp.; Volume II, 684 pp.; Volume III, 866 pp.

The U. S. government publishes two important series of documents on specific air pollutants. One series deals with "air quality criteria," which are established from our knowledge of the effects of air pollutants. The other series outlines "control techniques." Publication of these volumes started in January, 1969, under the National Air Pollution Control Administration, Washington, D. C. The agency now responsible for these publications is the Environmental Protection Administration. Examples of specific titles are

Air Quality Criteria for Sulfur Oxides.
Air Quality Criteria for Particulate Matter.
Control Techniques for Particulate Air Pollutants.

For a very brief introductory text, refer to

National Tuberculosis and Respiratory Disease Association. *Air Pollution Primer.* New York: 1969. 104 pp.

Various popular books that warn of the dangers of air pollution have appeared in recent years. Some representative ones follow:

L. J. Battan. *The Unclean Sky.* New York: Doubleday and Co., 1966. 141 pp.
D. E. Carr. *The Breath of Life.* New York: W. W. Norton and Co., 1965. 175 pp.
Howard R. Lewis. *With Every Breath You Take.* New York: Crown Publishers, 1965. 322 pp.

6

WATER
POLLUTION

6.1 THE NATURE OF WATER POLLUTION

The terms water pollution and air pollution both imply the presence of undesirable foreign matter in an otherwise "pure" or "natural" substance. However, the concept of pure water is quite different from that of pure air. As we learned in the preceding chapter, air is a mixture of several components, and "pure air" is therefore considered to be a particular mixture that represents a sort of ideal terrestrial atmosphere. Water, however, is a single compound, not a mixture. Therefore the chemist thinks of "pure water" as a substance consisting of molecules of only one type— those represented by the formula H_2O.* However, most drinking water contains small quantities of dissolved mineral salts; these substances often contribute to its taste. Thus one speaks of "pure spring water" in the sense of a natural mixture of water and small amounts of harmless, and perhaps tasty, mineral matter.

The pollution of water, then, is the addition of undesirable foreign matter which deteriorates the *quality* of the water. Water quality may be defined as its fitness for the beneficial uses which it has provided in the past—for drinking by man and animals, for the support of a wholesome marine life, for irrigation of the land, and for recreation. Pollutant foreign matter may be either non-living,

*We are ignoring differences in mass numbers that are due to the presence of different isotopes.

(Photo by Donald Chandler, courtesy of Dr. Arthur D. Hasler, Professor of Zoology, University of Wisconsin.)

such as compounds of lead or mercury, or living, such as micro-organisms. Again, there is an important difference between air and water. The pollution of air can usually be considered to be the discharge of foreign matter into empty space—the space not occupied by the air molecules. The air molecules neither invite nor resist foreign matter; they simply leave ample room for it.*

Consider a very different situation. It is not easy to pollute an iron bar, except on its surface. Immerse it in water or oil, expose it to bacteria or viruses, and its internal composition is unchanged. The reason for this resistance is that the atoms of iron are strongly and closely bonded to each other, and are very difficult to displace. This strong tendency for the atoms to retain their respective positions and not to be dislocated is what makes iron a solid. (For that

---

*We are not considering *secondary* air pollutants, which are formed by subsequent chemical reactions such as those of organic matter with oxygen and nitrogen under the influence of sunlight.

matter, it is not easy to pollute ice either; polluted ice is produced by the freezing of polluted water, not by the pollution of pure ice.)

Liquids are intermediate between gases and solids in their readiness to accept contamination. The attractive forces between molecules in the liquid state are strong enough that a sample of liquid (for example, a raindrop) holds itself together. However, the attractive forces are not so strong as they are in solids; they are not strong enough to prevent the molecules from sliding past one another. Such molecular relocations manifest themselves in the familiar phenomenon of liquid flow. Now, when a molecule of a liquid relocates, it leaves behind a vacant site, or a "hole." This vacancy can be occupied by another molecule of the same type, or by a molecule of a foreign substance.

What determines whether a given liquid will accept or reject a given type of foreign matter? We drop a lump of sugar into a glass of water and note that it dissolves. But a piece of lead does not. Since sugar is a solid, the sugar molecules must be attracted to each other—otherwise they would fly apart and sugar would be a gas. But the sugar molecules are also strongly attracted by water molecules, and can be pulled away from other sugar molecules to occupy sites surrounded by molecules of water. Thus, sugar dissolves in water. Atoms of lead are attracted to each other much more than they are to water molecules; thus, lead is insoluble. Therefore, the ease with which a liquid can become contaminated by dissolved foreign matter depends on the chemical relationships between the molecules of the liquid and foreign molecules.

A contaminant may be harbored by a liquid without being dissolved in it. If we grind our piece of lead to a fine powder and stir it into the water, the suspended lead is a pollutant. However, the ease with which foreign matter can be suspended in a liquid also depends to some extent on the mutual attraction between the foreign particles and the liquid molecules.

Water is not a typical liquid. Corn oil is more like cottonseed oil, kerosene is more like gasoline, and grain alcohol is more like wood alcohol than *anything* is like water. One of the consequences of the unique physical and chemical properties of water is that it invites or accepts pollution readily, sometimes through mechanisms that are quite unexpected. Of course, water is the universal liquid medium for living matter; it is therefore uniquely prone to pollution by living organisms, including those that carry disease to man. Contamination pathways that involve suspension, solution, and biochemical change are not necessarily separate and distinct from each other, and many of these complex processes can occur *only* in water.

Therefore, to understand water pollution, we must first consider the nature of water itself.

6.2   WATER

Wipe a drinking glass with a dry towel until it is sparklingly clean. Hold the glass upside-down just over a candle flame for five to ten seconds. The inside surface of the glass becomes clouded and wet. You have produced water from the hydrogen in the candle and the oxygen in the air. The oxygen in the air is oxygen gas, and the hydrogen in the candle is chemically bonded to carbon. The chemical composition of water can be established by weighing the ingredients that combine to produce it, although candle wax plus air is not the best choice for such an experiment. We find that water contains eight parts of oxygen to one part of hydrogen by weight. For example, nine pounds of water contains eight pounds of oxygen and one pound of hydrogen. The atomic weight of oxygen is 16, that of hydrogen is 1. Thus, an oxygen atom is 16 times as heavy as a hydrogen atom.* The composition of water, and the atomic weights of oxygen and hydrogen, are reconciled by the formula, H_2O: the composition of water is O (16) to H (1 + 1), or 16 to 2, which is 8 to 1. The sum of the atomic weights, $1 + 1 + 16 = 18$, is the molecular weight of water. The oxygen is bonded to the two hydrogens, and the molecule has a bent shape with an angle between the bonds of 105°.

$$\text{H} \diagdown \overset{+ \ 105° \ +}{\diagup} \text{H}$$
$$\text{O}$$

The negative charges (electrons) are crowded somewhat closer to the oxygen atom than the positive charges (protons) are. The effect of this is a separation of charges with the negatively charged part of the molecule nearer the oxygen atom and positively charged parts nearer the hydrogens. These electrical charges attract their opposites in other water molecules, with the result that liquid

---

*Strictly speaking, this ratio refers to the mass numbers of the more abundant isotopes. It is a good approximation, however, of the relative average atomic weights (see page 210).

water consists of aggregates of H_2O molecules bonded to each other as indicated below.

This strong aggregation accounts for the fact that water remains a liquid up to 100°C at normal pressure, in sharp contrast to the behavior of other substances of similar or even higher molecular weight. (For example, ether, molecular weight 74, boils at only 35°C.) The electrical forces that bind water molecules to each other can also serve to bind water molecules to those of foreign substances. Therefore, water is an unusually good solvent, especially for substances which have separated centers of positive and negative electrical charge. Such substances are, typically, inorganic compounds, such as the compounds of the metallic elements. Water is a poor solvent for substances whose molecules do not have separated centers of positive and negative charge; examples are the hydrocarbon substances derived from petroleum, such as gasoline, oil, and grease.

6.3   TYPES OF IMPURITIES IN WATER

It is useful to classify foreign substances in water according to the size of their particles, because this size often determines the effectiveness of various methods of purification. Figure 6.1 shows a spectrum of particles arbitrarily divided into three classes: suspended, colloidal, and dissolved. Let us consider each in turn as we refer to the figure.

Suspended particles are the largest—those which have diameters above about 1 micrometer. They are large enough to settle out at reasonable rates and to be retained by many common filters. They are also large enough to absorb light and thus make the water that they contaminate look cloudy or murky.

Colloidal particles are so small that their settling rate is insignificant, and they pass through the holes of most filter media; therefore they cannot be removed from water by settling or ordinary

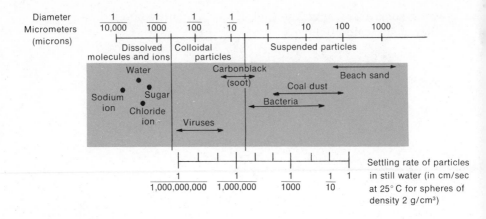

(25,400 microns = 1 inch)

Figure 6.1. Small particles in water.

filtration. Water that contains colloidal particles appears clear in the direct path of the light that illuminates it, but can look cloudy when observed at right angles to the light beam. (The same phenomenon occurs in air — colloidal dust particles can best be seen when observed at right angles to a sharply focussed light beam in an otherwise dark room.) The colors of natural waters, such as the blues, greens, and reds of lakes or seas, are contributed largely by colloidal particles.

Dissolved matter does not settle out, is not retained on filters, and does not make water cloudy, even when viewed at right angles to a beam of light. The particles of which such matter consists are no larger than about 1/1000 of a micrometer in diameter. If they are electrically neutral, they are called molecules. If they bear an electric charge, they are called ions. Cane sugar (sucrose), grain alcohol (ethanol), and "permanent" antifreeze (ethylene glycol) are substances that dissolve in water as electrically neutral molecules. Table salt (sodium chloride) on the other hand, dissolves as positive sodium and negative chloride ions.

Of course, foreign substances may be classified by properties other than particle size, for example, they may be living or non-living, organic or mineral, radioactive or non-radioactive, toxic or harmless, natural or added by man. It will be of most interest to us to focus our attention on those impurities that adversely affect the quality of water, to study their harmful effects and to consider the problems involved in removing them. Since natural water is never chemically pure, let us look first at impurities that occur in natural water, whether or not they are considered to be pollutants.

6.4 THE COMPOSITION OF NATURAL WATERS

Natural waters are never pure. The sources and natures of their impurities are outlined in Table 6.1. Of course, the proportions of these substances vary over wide ranges. Natural waters range in quality from tastily potable to poisonous; in saltiness, from fresh rainwater (non-salty) to brackish (partly salty, as where river water starts to mix with sea water), to ocean water, to the heavy concentrations of a landlocked evaporation sink such as the Dead Sea or the Great Salt Lake.

TABLE 6.1. IMPURITIES IN NATURAL WATERS

SOURCE	PARTICLE SIZE CLASSIFICATION				
	SUSPENDED	COLLOIDAL	DISSOLVED		
			Molecules	*Positive ions*	*Negative ions*
Atmosphere	←Dusts→		Carbon dioxide, CO_2 Sulfur dioxide, SO_2 Oxygen, O_2 Nitrogen, N_2	Hydrogen, H^+	Bicarbonate, HCO_3^- Sulfate, SO_4^{2-}
Mineral soil and rock	←Sand⟶ ←Clays⟶ ←Mineral soil particles→		Carbon dioxide, CO_2	Sodium, Na^+ Potassium, K^+ Calcium, Ca^{2+} Magnesium, Mg^{2+} Iron, Fe^{2+} Manganese, Mn^{2+}	Chloride, Cl^- Fluoride, F^- Sulfate, SO_4^{2-} Carbonate, CO_3^{2-} Bicarbonate, HCO_3^- Nitrate, NO_3^- Various phosphates
Living organisms and their decomposition products	Algae Diatoms Bacteria ←Organic soil (topsoil)→ Fish and other organisms	Viruses Organic coloring matter	Carbon dioxide, CO_2 Oxygen, O_2 Nitrogen, N_2 Hydrogen sulfide, H_2S Methane, CH_4 Various organic wastes, some of which produce odor and color	Hydrogen, H^+ Sodium, Na^+ Ammonium, NH_4^+	Chloride, Cl^- Bicarbonate, HCO_3^- Nitrate, NO_3^-

6.5 MICROORGANISMS IN WATER

Polluted water may be unsightly, odorous, corrosive, difficult to wash clothes in, or unpleasant to taste. But certainly, the most harmful effect of polluted water on man has been that of disease transmission. Typhoid fever in the Western Hemisphere and cholera in the Eastern Hemisphere have accounted for most of the waterborne deaths. Other human diseases carried to man by microorganisms in water are dysentery, infectious hepatitis, and gastroenteritis. Some viral diseases, such as poliomyelitis, may also be waterborne.

Man lives in intimate relationship with microorganisms on his skin and in his digestive system. The total microbial population in a normal human is perhaps 10 trillion (10^{13}), which is about a slimy cupful in volume. In health, humans and microbes live together to mutual benefit. The excrement of a healthy human, when dispersed in water, is not necessarily a source of disease. However, some healthy people live in harmony with organisms that can be pathogenic to others. For example, many Mexicans are adapted to waters carrying the bacilli that bring dysentery to tourists. Furthermore, not all humans (or animals) are healthy, and the presence in water of any microorganisms that are associated with sewage or excrement means that the water may also contain disease-carrying organisms. Such water must therefore be assumed to be unsafe. It is unrealistic, however, to demand that drinking water be absolutely free of intestinal bacteria. Such a standard would make drinking water too expensive, and it is not generally required for safety. In the United States, water is usually considered to be acceptable for drinking if (a) it contains fewer than 10 intestinal bacteria per liter; (b) it does not contain chemical impurities in concentrations that may be hazardous to the health of the consumer or corrosive to the water supply system; (c) it does not have any objectionable taste, odor, color, or cloudiness; and (d) it does not originate from a source that is subject to pollution from sewage or other contaminants, such as a well just downhill from a septic tank.

It is assumed that the number of disease-carrying microorganisms in water is proportional to the total number of microorganisms, and that a low total count therefore implies safety. There have been some cases, however, in which viral diseases have been transmitted by water that satisfies rigid bacterial standards. The presence of any impurities that are typical of sewage, even if they are not harmful in themselves, therefore implies that the water in which they occur is not protected against disease and is therefore unsafe. Abnormal concentrations of nitrogen compounds, such as ammonia (NH_3), or of chlorides serve as indices of the presence of these impurities.

6.6   NUTRIENTS AND OXYGEN IN WATER

Most organic matter derived from food wastes, from domestic sewage, and from plant residues such as soil particles, is decomposed in water by bacteria, protozoa, and various larger organisms. Such decompositions convert energy-rich substances to energy-poor ones by chemical reactions that utilize oxygen; this process is part of the food cycle described in Chapter 1. Of course, these conversions occur on land as well as in water. The important difference between the two environments, however, is that the atmospheric oxygen available to land animals is relatively rapidly replaced by plant life and thus does not become depleted — insects, birds, and mammals may compete for grain, but not for oxygen. The oxygen dissolved in waters, however, can be depleted faster than it is replaced from the atmosphere, and therefore bacteria, protozoa, sludge worms, and trout do compete for oxygen when organic nutrients are plentiful. Such competition affects the distribution of the forms of life in water. When the introduction of nutrient matter (such as garbage or organic industrial wastes) alters this distribution in a way that is unfavorable to man (for example, to discourage trout and favor protozoa), the quality of the water must be regarded as deteriorated, and the added nutrients are therefore pollutants.

When a fuel such as wood or candle wax or methane gas is burned in air, the result is the release of energy and the formation of the energy-poor substances carbon dioxide and water. Such a conversion is illustrated by the chemical equation for the combustion of methane, CH_4:

$$CH_4 + 2O_2 \rightarrow CO_2 + 2H_2O$$

Bacterial decomposition does not occur at flame temperatures; the specific sequences of alterations of chemical bonds are therefore quite different from those that take place in combustion. The overall release of energy, however, depends only on the starting materials and the final products, and not on the intermediate steps.[*] Therefore, it is valid to consider overall simplified chemical equa-

---

[*]This is one of the statements of the First Law of Thermodynamics. See Chapter 9.

tions in the study of the relationship between energy production and water pollution.

Bacterial decomposition in the presence of air is called **aerobiosis**, and it is the process that yields the most energy from a given weight of nutrients. For example, the complete aerobiosis of glucose (a sugar, $C_6H_{12}O_6$) may be represented by the equation

$$C_6H_{12}O_6 + 6O_2 \rightarrow 6CO_2 + 6H_2O + 3600 \text{ calories per gram of glucose}$$

Proteins contain sulfur and nitrogen as well as carbon, hydrogen, and oxygen. The aerobic decomposition of a protein is represented by the following unbalanced equation, in which the very complex protein molecule is represented only by a general formula, $C_xH_yO_zN_2S$:

$$C_xH_yO_zN_2S + O_2 \rightarrow CO_2 + H_2O + NH_4^+ + SO_4^{2-}$$
$$\text{ammonium} \quad \text{sulfate}$$
$$\text{ion} \quad \text{ion}$$
$$+ 5090 \text{ calories per gram of protein}$$

The reactions represented above are typical of the first stage of bacterial action in the deoxygenation of polluted waters. When organic nutrient is exhausted, additional energy can be obtained by the oxidation of ammonium salts.

$$NH_4^+ + 2O_2 \rightarrow 2H^+ + H_2O + NO_3^-$$
$$\text{ammonium} \quad \text{hydrogen} \quad \text{nitrate}$$
$$\text{ion} \quad \text{ion} \quad \text{ion}$$
$$+ 4350 \text{ calories per gram of ammonium ion}$$

This process is called **nitrification**.

What is most significant about these reactions from the viewpoint of water pollution is that they deplete the oxygen content of the water; oxygen molecules were required for each of the preceding reactions to take place. However, bacterial action does not stop when the molecular oxygen is gone. Instead, a new series of decompositions by the process called **anaerobiosis** occurs. The anaerobic decomposition of sugars and other carbohydrates is called **fermentation,** and that of proteins is called **putrefaction.** The latter process is represented by the following simplified unbalanced equation:

$$C_xH_yO_zN_2S + H_2O \rightarrow NH_4^+ + CO_2 + CH_4 + H_2S$$
$$\text{protein} \quad \text{methane} \quad \text{hydrogen}$$
$$\text{sulfide}$$
$$+ 368 \text{ calories per gram of protein}$$

Note that putrefaction yields much less energy (fewer calories) than oxidation, but it is still energetically profitable. Methane is very insoluble in water, and practically all of it is evolved as a gas. Hydrogen sulfide is highly odorous, resembling rotten eggs. Putrefaction, therefore, makes water bubble with foul smells and makes it unlivable for fish or other oxygen-breathing animals. It may be regarded as the worst condition of bacterial pollution.

We have seen that nutrient matter pollutes water because it serves as food for microorganisms. Microorganisms, including any pathogens that may be among them, multiply; the oxygen is exhausted and thus becomes unavailable for forms of life (such as fish) that man prefers; and finally, the stinks of putrefaction set in. What is the measure of such pollution? One might think that the analysis of a sample of water to determine the total amount of organic matter it contained would provide such an index. But not all organic matter is equally digestible by bacteria. In fact, some organic matter that is manufactured by industrial processes and is foreign to natural food chains may not be able to function as a nutrient at all. Such matter is said to be **non-biodegradable**. Some material, such as petroleum oil, may decompose only very slowly, so that it cannot be considered equivalent to, say, sugar as a nutri-

Figure 6.2. *"Notice how bright and white Brand X gets your clothing because of the harmful chemicals and enzymes it contains. Pure-O, on the other hand, containing no harmful ingredients, leaves your clothes lacklustre gray but protects your environment."* (Drawing by Dana Fradon; © 1971 The New Yorker Magazine, Inc.)

ient. An appropriate measure of pollution of water by organic nutrients is, therefore, an assay of the rate at which its nutrient matter can consume oxygen by bacterial decomposition. This assay is called the **biochemical oxygen demand, BOD**. Of course, the rate of biochemical oxidation depends on the temperature of the environment and on the particular kinds of microorganisms and nutrients present. If these factors are constant, then the rate of oxidation can be expressed in terms of the half-life of the nutrient. This concept is exactly the same as that applied to radioactive decay (see Chapter 4). The half-life is the time required for half of the nutrient to decompose, and the continuously decreasing rate can be represented in the form of a decay curve ("decay" here has both its biological and mathematical meanings) like that of Figure 4.3.

6.7 DETERGENTS, ALGAE, AND THE DEATH OF WATERS

Algae are aquatic plants; they are sometimes visible as a blue-green slime on the surface of still water. As plants, algae derive energy from photosynthesis. Therefore they consume carbon dioxide, CO_2, in the presence of sunlight and release oxygen. Like other plants, algae also need various inorganic nutrients, such as compounds of nitrogen, potassium, phosphorus, sulfur, and iron. The extent to which a body of water such as a lake can support algae depends on the inorganic nutrients that it can furnish; these nutrients become gradually more abundant as mineral matter is carried into the lake by streams. When the nutrient supply becomes sufficiently lavish, algae grow rapidly and can cover the surface of the water as thick, slimy mats. As some of the algae dies, either through exhaustion of some essential nutrients or for other reasons, it becomes, in turn, food for bacteria. As we have seen, bacterial decomposition consumes oxygen, with consequent polluting effects. Such a sequence of processes results in a condition in which the water beneath the slimy surface is deficient in oxygen and therefore unable to support forms of life that are useful to man. Perch and bass give way to less desirable scavenger varieties such as catfish, and to leeches and worms. A lake in such a condition is said to be **eutrophic**, and the process by which it is reached is called **eutrophication**. When it occurs without man's intervention, the process is usually slow, spanning hundreds or thousands of years, and is part of the overall evolution of a lake into a swamp and, eventually, into a meadow.

Man hastens the eutrophication of a lake whenever he adds plant nutrients to it. Agricultural fertilizers are plant nutrients; when they are applied to the soil some of them are carried away by flowing water and thereby hasten the aging of lakes.

The use of modern detergents has also contributed to the over-feeding of algae. To appreciate this somewhat complex situation, it will be helpful to understand something about the nature and mode of action of detergents. We mentioned earlier in this chapter that the ease with which a foreign substance can dissolve in a liquid depends on how strongly the molecules of the two different substances attract each other, relative to the mutual attractions of like molecules. Sugar dissolves in water, and if your hands are sticky with honey or lollipops they can be washed clean by rinsing them in pure water. Vegetable oil and animal fat, however, are insoluble in water, and the pure water that rinsed away the honey will leave any grease on your hands behind. It was known in ancient Rome that heating a mixture of animal fat and wood ashes produced a substance that could dissolve both in water and in grease and that could somehow bring these two otherwise incompatible substances together. Therefore, if you rinse your greasy hand with a mixture of water and this new substance, which we now call **soap**, the grease can be washed away. Soap functions in this manner because it is made up of long molecules that have, on one end, separated regions of positive and negative electrical charge that are strongly attracted to water molecules, and, on the other end, a hydrocarbon character that is attracted to grease molecules. This action of the soap molecule is called **detergency**.

The use of soap has not contributed to the eutrophication of lakes. Soap is a nutrient for bacteria, not for plants, and it is normally degraded by bacterial action in sewage. However, soap has not been a satisfactory detergent in all respects. The mineral matter in ground water contains metal ions that make soap insoluble, and thereby rob it of its detergency. This insoluble soap manifests itself as a "ring around the bathtub" or a "tattletale gray" on an otherwise white textile fabric. Water that contains such mineral matter is called "hard water." Water from which it is absent, such as rain, is said to be "soft." The years since World War II have witnessed increasing development and use of synthetic detergents that are effective in hard water and that have various other properties that offer advantages over soap. However, these synthetic

A

B

Figure 6.3. The choking of waters by weeds. *A*. The dam on the white Nile at Jebel Aulia near Khartoum, Sudan. The area was clean when photographed in October 1958. *B*. The same area in October 1965, showing the accumulation of water hyacinth above the dam. (From Holm: Aquatic Weeds. *Science 166*:699–709 (Nov. 7) 1969. Copyright 1969 by the American Association for the Advancement of Science.)

detergents contain plant nutrients and thereby hasten eutrophication. Phosphates* are frequently in short supply in natural waters, and their replenishment from detergents has therefore been implicated as being particularly responsible for the increased nourishment of algae and the consequent eutrophic deterioration of lakes.

It is important to recognize that waterways are frequently delicately balanced ecosystems, and can be disrupted by pollutants to a degree that far exceeds unsophisticated expectations. In many areas of the world, especially in the great rivers and lakes of the tropical and subtropical regions, aquatic weeds have multiplied explosively. They have interfered with fishing, navigation, irrigation, and the production of hydroelectric power. They have brought disease and starvation to communities that depended on these bodies of water. Water hyacinth in the Congo, Nile, and Mississippi rivers and in other waters in India, West Pakistan, Southeast Asia and the Philippines, the water fern in southern Africa, and water lettuce in Ghana are a few examples of such catastrophic infestations. Man has always loved the water's edge. To destroy the quality of these limited areas of the earth is to detract from his humanity as well as from the resources that sustain him.

6.8   INDUSTRIAL WASTES IN WATER

Industrial activity, especially pulp and paper production, food processing, and chemical manufacturing, generates a wide variety of waste products that may be discharged into flowing waters. Some of these wastes are known to be poisonous to man; the effects of others are obscure. Some have been known since antiquity; many are quite recent, and new types of wastes continue to appear as new technology develops. Many industrial wastes are organic compounds that can be degraded by bacteria, but only very slowly, so that they may carry unpleasant odors and tastes along a watercourse for considerable distances. (Even domestic sewage contains significant quantities of non-biodegradable substances of unknown

---

*Phosphates are ions that contain the elements phosphorus, oxygen, and perhaps hydrogen. Examples:

Orthophosphate, PO_4^{3-}
Pyrophosphate, $P_2O_7^{4-}$
Metaphosphate, PO_3^-
Monohydrogen phosphate, HPO_4^{2-}
Dihydrogen phosphate, $H_2PO_4^-$

In 1967, the average synthetic detergent manufactured in the United States contained about 9.4% phosphorus. In 1969, the phosphorus content of "enzyme presoaks" ranged from 15% to 17%.

origin.) To complicate matters still further, some of these wastes react with the chlorine that is used as a disinfectant for drinking water. The result of such reaction is the production of chlorinated organic compounds that smell and taste much worse than the original waste product.

One of the oldest known waterborne industrial poisons is lead. Throughout history, the most prevalent source has been the lead piping formerly used in water distribution networks. More recently, the use of lead arsenate spray as an insecticide has contaminated surface and ground waters with both lead and arsenic. Lead is a cumulative poison, and even small concentrations, if continuously present in drinking water, may lead to serious illness or death. Arsenic, which sometimes occurs in natural waters that flow through arsenic-bearing minerals, is also a cumulative poison. The "safe" limits in drinking water for both lead and arsenic are recommended to be no higher than about 0.01 ppm.

The compounds of various other metals, such as copper, cadmium, chromium, and silver, have sometimes been implicated as industrial water pollutants. In very recent years, most attention has been given to the problem of mercury poisoning, and we shall devote the remainder of this section to that topic as an example of metallic pollution. Mercury has always been regarded with fascination and alarm. It is the only metal that is liquid at ordinary temperatures (hence its other name, quicksilver), and it is fun to play with. (But don't do it. Its vapor is poisonous, and at high temperatures it can vaporize rapidly enough to be deadly.) Some of its compounds, whose toxicity has been well known since the Middle Ages, have been used as agents of murder and suicide. Until very recently, however, mercury was not considered a dangerous water pollutant, for the following reasons: Although mercury is widely distributed over the earth, it generally occurs only in trace concentrations. Natural waters typically contain only a few parts per billion of mercury. Metallic mercury, itself, although poisonous in vapor form, is not particularly hazardous when taken by mouth as a liquid. The use of mercury as a component of dental fillings has been shown to be harmless; the mercury in the teeth does not migrate to other parts of the body. Many mercury compounds are very highly insoluble; for example, it is calculated that it would require about 25 gallons of water to dissolve one molecule of mercuric sulfide, HgS!

These considerations imply that mercury in water is not a potential pollutant and perhaps account for the previous lack of concern over the fact that half of the total amount of mercury mined annually is released into the environment. (About 10,000 tons are mined, of which 5000 tons are somehow "lost.") These discharges occur as waste effluents from manufacturing plants or else as the incorporation of traces of mercury into products in which it does not belong. An example is the electrochemical conversion of brine, NaCl dissolved in water, into chlorine and sodium hydroxide, as represented by the equation:

$$2NaCl + 2H_2O \rightarrow Cl_2 + 2NaOH + H_2$$

| | chlorine | sodium hydroxide (caustic soda) | hydrogen, which is released to the atmosphere |

Mercury does not appear in the equation, but it flows along the bottom of the reaction cell as an electrical boundary (electrode) at which the sodium hydroxide and hydrogen are produced. When the salt solution (brine) becomes too weak, it is discarded. This waste contains mercury, which then follows whatever watercourse is available to it. The sodium hydroxide product also is contaminated with mercury, and carries it into many products for which sodium hydroxide is a raw material. Finally, the hydrogen discharged to the atmosphere also carries some mercury vapor with it.

The complacent notion that such discharges are tolerable has been destroyed by various instances of acute mercury poisoning. Most notable was that which occurred in the 1950's in a coastal area of Japan known as Minamata Bay, where fishermen, their families, and their household cats all became stricken with a mysterious disease that weakened their muscles, impaired their vision, led to mental retardation, and sometimes resulted in paralysis and death. What the people and their cats had in common was a diet of fish, and what the fish had in their bodies was a high concentration of mercury that came from the bay waters. The Minamata Bay received the mercury-containing effluent from a local plastics factory. Moreover, the mercury in the fish was present in organically-bound forms that are especially hazardous to man. These compounds are all related to methyl mercury, $H_3C \cdot Hg \cdot CH_3$. Such mercury compounds are sometimes used as pesticides and fungicides, and the discharge of these residues into waters is therefore a serious hazard. Following this episode, research led to the disturbing finding that metallic mercury and inorganic mercury compounds can be methylated (converted to methyl mercury) by anaerobic bacteria in the mud of lake bottoms, as well as by fish and mammals. The mercurial wastes that have accumulated in muddy lake bottoms therefore cannot be regarded as inert sludges; they are potential sources for biochemical conversion into forms of mercury that can enter and pass through the food chain in increasing concentrations and thus become poisons for man.

It would be easy to draw the conclusion from these disturbing circumstances that it would be good to "abolish" mercury, if that were somehow possible. That conclusion would be incorrect. Mercury has been present in the environment, including the food chains, throughout the history of life on Earth, and man has necessarily developed a tolerance for the concentrations to which he has been exposed during his evolution. "Tolerance" is not a passive detachment; it is a biochemical adjustment, and such an adjustment usually leads to dependence. It is therefore likely that man requires small amounts of mercury, as he does traces of other metallic elements that would be poisonous in higher concentrations. To say that mercury is the danger is therefore only a limited truth; the more general conclusion is that the danger lies in heedless disruption of a delicately balanced ecosystem.

6.9   CORROSIVENESS

Acids corrode (rust, oxidize) metals, and, as we have seen, soluble compounds of metals can pollute water. Therefore water that is acidic can become contaminated with metallic pollutants more easily than pure water can.

The original meaning of acid is "sour," referring to the taste of substances such as vinegar, lemon juice, unripe apples, and old milk. It was observed that all these sour or acidic substances had some properties in common: Their presence altered various natural vegetable colors; for example, lemon juice lightens the color of tea. It was also observed that the corrosion of metals by acids was accompanied by the release of hydrogen. The presence of hydrogen is, in fact, essential to the quality of acidity in water. The hydrogen must be in the form of positively charged hydrogen atoms that are called hydrogen ions, H^+, which are attached to water molecules. Hydrogen ion concentration is measured in terms of an expression called pH (for "power of hydrogen"). The relationship between the two units is shown in the following table:

Concentration of H⁺ Ion (Grams/Liter)	pH
$\dfrac{1}{10}$	1
$\dfrac{1}{100}$	2
$\dfrac{1}{1000}$	3
$\dfrac{1}{1,000,000}$	6
$\dfrac{1}{10,000,000}$	7

The pH value is thus the number of zeros in the denominator of a fraction that expresses H^+ ion concentration, whose numerator is 1 and whose denominator is a multiple of ten.*

Pure water at 25° C has an H^+ concentration of 1/10,000,000, or a pH of 7, and is said to be "neutral". Any pH less than 7 connotes acidity. The lower the pH of a body of water, the more prone it is to be corrosive and, thereby, to become polluted with metallic compounds. Acidic waters coursing through lead pipes will therefore become more toxic than pure water. And grapefruit juice standing in an iron cup develops a terrible taste.

6.10   WATER PURIFICATION

The purification of water has developed into an elaborate and sophisticated technology, much of which is beyond the scope of this book. However, the general approaches to purification should be comprehensible, and in some cases even obvious, from a general understanding of the nature of water pollution. If water contains impurities that will settle out, hold the water still long enough for settling to occur, or filter them if they can be retained on a filter.

---

*This discussion of pH is highly simplified. For a more elaborate treatment, see any standard chemistry textbook. pH values are actually logarithmic expressions, and an explanation of a related log scale, the decibel scale of sound, appears in Chapter 10.

If the particles are too small for either process, make them larger by causing them to stick together, or coagulate, in some way, so that settling or filtration becomes feasible. If the water is so acidic that it is corrosive, neutralize the acid. Oxidize the organic waste. Kill the microorganisms. Remove the bad tastes and odors with some appropriate agent, such as activated carbon. "Soften" the hard water so that non-polluting detergents can be used effectively.

Some insight into the workings of water purification systems will be gained by tracing the course of a typical process: We will examine the method of sewage treatment.

Figure 6.4. Secondary and tertiary sewage treatment plant of the South Tahoe Public Utility District in 1965. Since this photograph was taken, the capacity of this plant has been increased from 2.5 to 7.5 million gallons per day and even more advanced waste treatment has been provided. This has been necessary to prevent eutrophication of Lake Tahoe, high in the Sierra Nevada Mountains lying between California and Nevada. Attracted by this beautiful lake, vacationers and permanent residents have increased in number many times in recent years. The plant, as shown before expansion, provided for primary sedimentation (building in center right), activated sludge treatment (lower right), secondary sedimentation (circular pond), sludge digestion (large circular tanks), and tertiary treatment (building in upper right). (From Warren: *Biology and Water Pollution Control.* Philadelphia: W. B. Saunders Co., 1971. Photo courtesy of Cornell, Howland, Hayes, and Merryfield, Engineers and Planners, Corvallis, Oregon.)

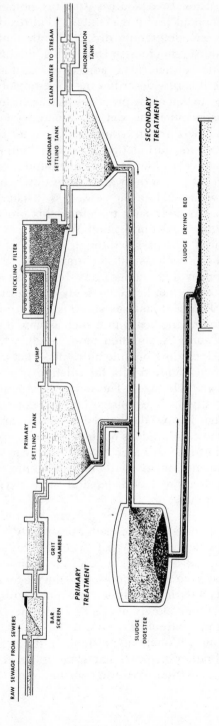

Figure 6.5. Sewage plant schematic, showing facilities for primary and secondary treatment. (From *The Living Waters*. U.S. Public Health Service Publication No. 382.)

Waterborne wastes from sources such as homes, hospitals, schools, and commercial buildings contain food residues, human excrement, paper, soap, detergents, dirt, cloth, other miscellaneous debris, and, of course, microorganisms. This mixture is called **sanitary** or **domestic sewage.** (The adjective "sanitary" is rather inappropriate since it hardly describes the condition of the sewage; it presumably refers to that of the premises whose wastes have been carried away.) Waste-containing waters, sometimes joined by the run-off from rain, flow through a network of street sewer pipes. Microbial action occurs during this flow; the high-energy food chemicals are degraded to low-energy compounds, with the consumption of oxygen. It is as if the waters were slowly burning. The more such action occurs before the sewage is discharged to open waters, the less occurs afterward; therefore, this process must be regarded as the beginning of purification. When the sewage reaches the treatment plant, it first passes through a series of screens that remove large objects, such as rats or grapefruits, then through a grinding mechanism that reduces any remaining objects to a size small enough to be handled effectively during the remaining treatment period. The next stage is a series of settling chambers designed to remove first the heavy grit, such as sand that rain water brings in from road surfaces, and then, more slowly, whatever other suspended solids — including organic nutrients — that can settle out in an hour or so. (See Figure 6.1 for the relationship between settling velocity and particle size.) The entire treatment up to this point has been relatively inexpensive but has not accomplished much. If the sewage is now discharged into a stream (as, unfortunately, is often the case), it does not look so bad because it bears no visible solids, but it is still a potent pollutant carrying a heavy load of microorganisms, many of them pathogenic, and considerable organic nutrients that will demand more oxygen as their decomposition continues.

The next step in the treatment is designed to reduce the dissolved or finely suspended organic matter drastically by some form of accelerated biological action. What is needed for such decomposition is oxygen and organisms and an environment in which both have ready access to the nutrients. One device for accomplishing this objective is the trickling filter, shown in Figure 6.6. In this device, long pipes rotate slowly over a bed of stones, distributing the polluted water in continuous sprays. As the water trickles over and around the stones, it offers its nutrients in the presence of air to an abundance of rather unappetizing forms of life. A fast-moving food chain sets in operation. Bacteria consume molecules of pro-

Figure 6.6. Picture of a trickling filter with a section removed so as to show construction details. (From Warren: *Biology and Water Pollution Control.* Philadelphia: W. B. Saunders Co., 1971. Photo courtesy of Link-Belt/FMC.)

tein, fat, and carbohydrate. Protozoa consume bacteria. Farther up the chain are worms, snails, flies and spiders. Each life form plays its part in converting high-energy chemicals to low-energy ones. It is as if the slow fire in the sewer were burning more brightly here. All the oxygen consumed at this stage represents oxygen that will not be needed later when the sewage is discharged to open water. Therefore, this process constitutes a very significant purification. The effluent from this biological treatment, however, may be "burned-out," and its demand for oxygen sufficiently reduced, but it is not free of microorganisms and hence may still carry disease. Since the microorganisms have done their work, they may now be killed. The final step is therefore a disinfection process, usually chlorination.

Of course, there are complications. Some industrial organic wastes serve poorly as nutrients, and are not only difficult to decompose, but may poison bacteria and thus interfere with the biochemical action on otherwise degradable materials. Non-degradable detergents cannot distinguish between the sewage treatment plant and the home washing machine, and continue to produce a most unwelcome foam.

Furthermore, it is not always advantageous to oxidize *all* the nutrients. After all, nutrients may be used to feed forms of life more appetizing than worms and flies. Properly managed with regard to oxygen, nutrients may be discharged into ponds stocked with edible fish such as carp or trout. Thus at least some of the waste from the dinner table may eventually be recycled to the dinner table.

PROBLEMS

1. Define water pollution; water quality.

2. Discuss the relative ease of contamination of gases, liquids, and solids.

3. Define molecule; ion; colloidal particle; suspended particle.

4. A healthy person lives in harmony with bacteria in his digestive system. Why, then, should water that contains digestive bacteria be considered to be polluted?

5. What are the criteria for water that is considered fit for drinking? Is such water always safe? Is water that does not meet these criteria always harmful? Explain.

6. Explain how a non-toxic organic substance, such as chicken soup, can be a water pollutant.

7. Define the terms aerobiosis; anaerobiosis; nitrification; fermentation; putrefaction.

8. What harmful effects on water quality result from the depletion of molecular oxygen?

9. Define biodegradability; biochemical oxygen demand. In what way is the latter a measure of water pollution?

10. What is eutrophication? Explain how it occurs and why it is hastened by the addition of inorganic matter such as phosphates.

11. List seven metals whose compounds may have been implicated as water pollutants.

12. The poisonous properties of mercury have been known since the late years of the Roman Empire, when mercury miners in Spain suffered chronic or acute symptoms, such as tremors and loss of hair and teeth, from exposure to mercury vapor. In more recent years, hatters exhibited the same symptoms (which they called "the hatter's shakes") from the mercury compounds used to convert fur to felt. Do you think these circumstances should

necessarily have led to the conclusion that mercury would be a water pollutant if the metal were discharged into streams or lakes? What facts actually led to the opposite conclusion? Is mercury dangerous as a water pollutant in the same manner that it is in mines or hat factories? Explain.

13. The contents of our stomachs are acidic, and we drink acidic fruit juices without doing ourselves any harm. Why, then, are acids considered to be pollutants in drinking water?

14. The following advice has been offered to tourists who wish to avoid ill effects from drinking water in areas where sanitation is uncertain or where intestinal disorders are common: Do not drink cold tap water. *Never* drink water from pitchers, carafes, or bottles which may have been reused, even if washed. On the other hand, tap water that is hot enough to burn your hand may be directly used to rinse and fill a cup and, when it has cooled, is safe to drink. Suggest a rational justification for each of these statements.

The following questions require arithmetic reasoning.

15. Is the speed of settling of particles in water directly proportional to their diameters? If the diameter is multiplied by 10, is the settling speed 10 times faster? (Justify your answer with data from Figure 6.1) Is a settling pond a good general method of water pollution control? Explain.

16. If the atomic weight of oxygen were 8 instead of 16, what would be the chemical formula for water?

Answer

16. HO

BIBLIOGRAPHY

There is a considerable volume of literature on water, on the analysis of its impurities, and on various aspects of its purification, including sewage treatment. A basic text on the properties of water itself is

Ernest Dorsey. *Properties of Ordinary Water-Substance.* New York: Litton Educational Publisher, Van Nostrand—Reinhold Books, 1940. 704 pp.

The following texts are good sources of information on water pollution and its control:

Charles E. Warren. *Biology and Water Pollution Control.* Philadelphia: W. B. Saunders Co., 1971. 434 pp.

T. R. Camp. *Water and its Impurities.* New York: Litton Educational Publisher, Van Nostrand—Reinhold Books, 1963.

G. V. James and F.T.K. Pentelow. 3rd ed. *Water Treatment.* London: Technical Press, 1963.

The following are citied as representative of popular books that deal with the crisis of water pollution:

D. E. Carr. *Death of the Sweet Waters.* New York: W. W. Norton and Co., 1966. 257 pp.

F. E. Moss. *The Water Crisis.* New York: Encyclopaedia Britannica, Praeger Publisher, 1967. 305 pp.

G. A. Nikolaieff, ed. *The Water Crisis.* Bronx, New York: H. W. Wilson Co., 1969. 192 pp.

7

▌ SOLID WASTES

7.1   SOURCES AND CYCLES

As we mentioned in Chapter 1, the mass of the Earth is relatively constant. The matter gained from the regions beyond our atmosphere (such as meteorites), and the matter lost (such as molecules from the upper atmosphere and, recently, space vehicles) is entirely negligible compared with the mass of the Earth or even with the mass of the biosphere — that part of the Earth where life occurs. However, the materials of the Earth are not stagnant: there are winds, ocean currents, rivers, evaporation, rainfall and snowfall, the creep of glaciers, and the various movements of matter in which life is involved.

In terms of time intervals that approximate the life spans of large animals (decades) or of large plant (centuries), these natural movements of matter are balanced and cyclic. Thus, for example, evaporation balances precipitation and the ocean levels remain constant. The balanced nature of food cycles was discussed in Chapter 1. Of course, such balance is not universal: volcanic eruptions, for example, are strictly one-way affairs. And long-term effects, such as continental drifts or upheavals or ice ages, do not occur as cycles within the life span of any organism.

Man's factories consume raw material to manufacture products which are eventually discarded as wastes. In Chapter 1 we pointed out that the waste products of living organisms are consumed as raw materials by other organisms. If this were not so, the waste products would accumulate ceaselessly, with the destruction of the ecosystem resulting. However, processes in factories differ from

135

those in living plants or animals; hence the matter that passes through factories follows new pathways. These pathways do not, in general, involve recycling. For example, coke, which is produced from coal, is used as a raw material for manufacturing the gas acetylene, which in turn is used for making various plastics and synthetic rubber. The plastics and rubber eventually accumulate in some location such as a garbage dump; they do not return to the mine as coal. In fact, many new synthetic materials, particularly plastics and corrosion-resistant coatings for metals, were developed to be resistant to chemical changes so that they would not deteriorate during their useful lifetimes. Unfortunately, this resistance also persists after the products are discarded. The movement of matter through the industrial processes, unlike the movement through the life processes, therefore generates an ever-increasing quantity of waste, mostly in the form of solid material. This does not mean that *every* industrial product eventually becomes a dead-end waste. Some products are used as raw material for other manufacturing. Other industrial products—for example, soap—can be used as foods by living organisms. As previously stated, materials that can be consumed by living organisms are **biodegradable**. However, the fact that a waste product is biodegradable does not necessarily mean that it is benign to the ecosystem in which it is discarded. For example, although petroleum is degraded by bacteria, the process is very slow. Tarry residues dumped along a shoreline may disrupt a particular ecosystem beyond recovery long before bacteria consume the tar.

In this chapter we shall discuss the sources of solid wastes, the extent to which they are recycled, and the problems and issues involved in their disposal. We exclude radioactive wastes, which were discussed in Chapter 4.

7.2 THE SOURCES OF SOLID WASTES

Perhaps the most noteworthy characteristic of solid wastes is their variety. We are familiar with the types of waste in our household garbage pails. In addition, there is combustible rubbish such as paper, cardboard, wood or leaves; there is non-combustible rubbish such as glass, bottles, crockery or cans, and furnace ashes and cinders; and there are large objects such as discarded automobiles, furniture, appliances, and rugs. However, household sources contribute only a small portion of the total solid wastes discarded in the United States, and constitute only a fraction of the varieties. Table 7.1 is a classification of industrial sources compiled

TABLE 7.1. MAJOR INDUSTRIAL SOLID WASTE CATEGORIES

Acetylene wastes	Manganese
Agricultural wastes	Mica
Aluminum	Mineral wool
Animal-product residues	Molasses
Antimony	Molybdenum
Asbestos	Municipal wastes
Ash, cinders, and flue dust	Nonferrous scrap
Asphalt	Nuts
Bauxite residue	Nylon
Beryllium	Organic wastes
Bismuth	Paint
Brass	Paper
Brewing, distilling, and fermenting wastes	Petroleum residues
Brick plant waste	Photographic paper
Bronze	Pickle liquor
Cadmium	Plastic
Calcium	Poppy
Carbides	Pottery wastes
Carbonaceous shales	Precious metals
Chemical wastes	Pulp and paper
Chromium	Pyrite cinders and tailings
Cinders	Refractory
Coal	Rice
Cobalt	Rubber
Coffee	Salt skimmings
Coke-oven gas residues	Sand
Copper	Seafood
Cotton	Shingles
Dairy wastes	Sisal
Diamond grinding-wheel dust	Slag
Distilling wastes	Sodium
Electroplating residues	Starch
Fermenting wastes	Stone spalls (chips)
Fish	Sugar beets
Flue dust	Sugar cane fibers
Fluorine wastes	Sulfur
Fly ash	Tantalum
Food processing wastes	Tetraethyl lead
Foundry wastes	Textiles
Fruit wastes	Tin
Furniture	Titanium
Germanium	Tobacco
Glass	Tungsten
Glass wool	Uranium
Gypsum	Vanadium
Hemp	Vegetable wastes
Hydrogen fluoride slag	Waste paper
Inorganic residues	Wood wastes
Iron	Wool
Lead	Yttrium
Leather fabricating and tannery wastes	Zinc
Leaves	Zircaloy
Lime	Zirconium
Magnesium	

by the Bureau of Solid Waste Management of the U.S. Department of Health, Education, and Welfare. Some of these waste materials are more or less biodegradable, some are combustible, some are toxic, some have foul odors, and some are inert, but all of them occupy space.

7.3 METHODS OF DISPOSAL

There are two kinds of pathways possible for solid waste materials: they may be recycled into other processes or they may accumulate somewhere. Of course, a given waste source may be partly recycled and partly accumulated. Non-returnable soft drink bottles do not recycle, so they must all accumulate somewhere, not necessarily together. Deposit bottles, on the other hand, do recycle, but not completely; some are never returned because they are broken, lost, or neglected. The average number of round trips that a deposit bottle used to make was about 30, but the indifference of the consumer to a two- or five-cent deposit reduced this to about four round trips before the shift to non-returnable bottles. The latter rate means that the "average" bottle will make five trips out of the bottling factory and will return four times, not to be seen again after the fifth departure. Thus, if a large number of bottles with random histories leaves the bottling factory at a given time, four out of every five will be destined for a round trip, and one for the scrap heap. Therefore the fraction being recycled is four-fifths, or 80%, still a fairly high proportion.

The total quantities of solid wastes are large and increasing. The average in the United States was 2.75 pounds per capita per day in 1920 and 4.5 pounds in 1965. In urban areas the average is much greater, reaching as high as eight pounds per day in some cities.

The scrap automobile is a particularly vexing problem; what has happened in this realm in recent years is somewhat analogous to what occurred in the shift from deposit to one-way bottles. Until a few years ago, the value of the scrap steel and other parts of an old car was great enough to induce its owner to sell it to a junkyard. However, because of changes in steel production and economics which have reduced the need for scrap, it now frequently costs more to move a car to the junkyard than it is worth. It is as if the car, like the soda bottle, were stamped "no deposit, no return." This situation has resulted in an increasing rate of abandonment of old cars on city streets. The total for New York City in 1969, for example, was 57,000.

*"Admit it. Now that they're starting to recycle
this stuff, aren't you glad I didn't throw it out?"*

Figure 7.1. (From Saturday Review of Literature, July 3, 1971. By
Joseph Farris.)

Figure 7.2. *Environment* — Robert Charles Smith.

Where do the solid wastes go? Sometimes we seem to feel that there is junk all over the landscape. But the distribution is far from random. Although some is dumped into bodies of water, solid wastes are amassed for the most part on the land, usually after some preliminary processing and concentration.

LAND DISPOSAL

The most primitive waste repository is the open dump. The operation is conceptually rather simple. Waste is collected and, to save space and transportation costs, is compacted. Compaction is effected by a householder when he flattens a can or squeezes a garbage bag, or by equipment especially designed for individual or multiple dwellings, or by packer-type refuse trucks, used in the United States since 1950, that reduce the material to as little as one-third of its original volume. The compacted waste is hauled to the dumping site, usually in the morning, and spread on the ground, further compaction sometimes being effected by bull-dozing. Organic matter rots or is consumed by insects, by rats, or, if permitted, by hogs. Various salvaging operations may go on during the day. Bottles, rags, knick-knacks, and especially metal scraps are collected by junk dealers or by individuals for their own use. In some communities, the accumulation is set afire in the evening (or it may ignite spontaneously) to reduce the total volume and to expose more metal scrap for possible salvage. Of course, the organic degradation, the burning, and the salvaging are recycling operations. However, there are serious detrimental features to the open dump. The biological environment differs from those devised by man (agriculture and animal husbandry) and from those that have evolved in natural ecosystems, and is therefore not controlled by regulatory mechanisms common to either. The result is that the organisms that multiply at the dump are not likely to be the type that are benign to man. In other words, the dump is a potential source of diseases, especially those carried by flies and rats. The fires, too, are uncontrolled and therefore always smoky and polluting. Rainfall that circulates enters the dump and removes a quantity of dissolved and suspended matter, including pathogenic microorganisms, that are water pollutants. And, of course, the dumps are ugly.

A more advantageous method of land disposal is the **sanitary landfill**, in which each layer of waste is covered with a layer of soil, clay, or gravel. For efficient operation, the waste must be well compacted, and large objects (such as furniture) must be shredded. In this way, the waste is not exposed to air, vermin, or rodents, but it is subject to bacterial decomposition, so that biodegradation takes place while pollution, disease, and unsightliness are avoided. In practice, the distinction between the sanitary landfill and the open dump is not always sharp. For example, a thin earth layer may be an ineffective barrier against burrowing rats, flies emerging from larvae, gases evolving from decomposition, or pollutants dissolving into the water.

There are two long-range aspects of land disposal that merit attention. One is that such disposal represents an effective loss of certain non-renewable supplies of metals, especially copper, zinc, and lead. The second must certainly be the most obvious: the character of the land is changed. In some cases this has been counted a benefit: for example, where marsh land is filled in and new areas become available for residential, commercial, or industrial development. However, such conversion is not an unmitigated gain; marsh lands are essential for the support of many varieties of bird and aquatic life, and their loss leads to serious ecological disruptions.

When marsh lands are not available, other low areas are usually used as dumping sites, because less work is required to fill a hole than to build up a hill. But the loss of low areas sometimes disrupts patterns of water run-off, resulting in increased water pollution and risks of flooding. It is generally ecologically more satisfactory to avoid such problems by building up mounds of waste on areas that are already high enough that they are not essential parts of natural watercourses. A rather appropriate name assigned to one such growing feature of the landscape is "Mount Trashmore."

INCINERATION

A method increasingly used in many metropolitan areas is incineration. The process, as applied to waste disposal, is more complex than simply setting fire to a mass of garbage in an open dump. There may be as many as four aspects to incineration. The first is the burning of the waste material itself; the chamber in which this occurs is called the furnace. The second involves the disposal of the residue — the ashes or "clinker." Third is the control of pollutants from the burning process. This can be effected by a secondary combustion chamber to complete the oxidation of all unburned gases coming from the furnace and by other devices to remove airborne particulate matter before the exhaust is released to the atmosphere. Finally, the heat may be recovered for some useful purpose, such as generation of process steam or electric power. The latter aspect sounds attractive — the use of garbage to generate electricity. Such recovery has been practiced in Europe for many years but not in the United States, where the cost trade-offs with older methods of power generation have been considered unfavorable. However, the potential heat content of our waste is increasing because more and more dry cellulose (such as paper and cardboard) is being discarded. This factor, taken together with that of increased land costs, may tend to bring the waste heat recovery operations into favor.

The incineration process has the following advantages: (a) it eliminates the health problem associated with refuse accumulation; (b) it reduces the volume of solid waste by about 80% and therefore requires much less land for eventual disposal of its residue; (c) it can handle a mixture of garbage and rubbish without prior separation; (d) equipment in a wide range of sizes, from apartment house units to large central municipal incinerators that handle over 1000 tons per day, can be used; (e) the residue (clinker) is inert and odorless and relatively easy to handle. However, incineration is disadvantageous in that it squanders raw materials just as dumping does.

7.4   RECYCLING

The discussions in the preceding section showed that accumulation and recycling are not always mutually exclusive. Rotting and burning do serve to recycle some wastes, but not all, so the dumps continue to grow. Some processes, however, involve essentially total recycling. Notable examples are composting, rendering, pyrolysis, and industrial salvaging.

Composting is the controlled, accelerated biodegradation of moist organic matter to a humus-like product that can be used as fertilizer or soil conditioner.

Rendering is the cooking of animal wastes, such as fat, bones, feathers, and blood, to yield both a fatty product called tallow, which is the raw material for soap, and a non-fatty product that is high in protein and can be used as an ingredient of animal feed.

Destructive distillation, or **pyrolysis**, is the process by which a material is decomposed by heating it in the absence of air. It has been found that valuable products may be recovered by the pyrolysis of municipal wastes (Figure 7.3), and this approach is being adopted in some locations. An added advantage is that the pyrolysis equipment is essentially a closed system and therefore does not discharge pollutants to the atmosphere.

Industrial salvaging comprises a very wide variety of processes; the common objective, however, is to recycle waste materials into manufacturing processes. The salvaging of metal wastes has the important added effect of conserving non-renewable resources.

It is easy to think that recycling operations are less disruptive of the Earth's ecosystem and therefore always more welcome than non-recycling operations. But the decisions do not always favor recycling. The reason is that not all of the people who may be affected by the necessary trade-off participate in the decision. As an example, consider the rendering process briefly described above. The raw material for a rendering plant comprises wastes from a wide variety of sources — farms, slaughterhouses, retail butcher shops, fish processing plants, poultry plants, and canneries. If there were no rendering plants, these wastes would impose a heavy burden on sewage treatment plants, as well as add pollutants to streams and lakes and nourish disease organisms. At the rendering plant, the waste materials are sterilized and converted to useful products, such as tallow and chicken feed. But the cooking process generates odors. These odors can be controlled by methods such as incineration, but any control method is subject to occasional interruptions, such as those that might be caused by a power failure.

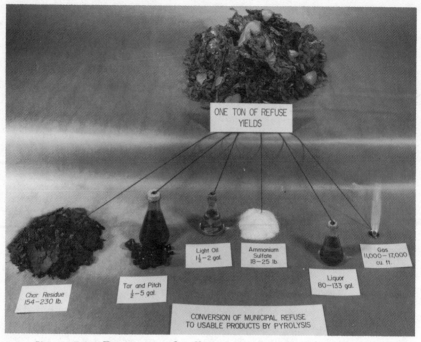

ONE TON OF REFUSE
YIELDS

Light Oil
1½-2 gal.

Ammonium
Sulfate
18-25 lb.

Gas
11,000-17,000
cu. ft.

Liquor
80-133 gal.

Char Residue
154-230 lb.

Tar and Pitch
½-5 gal.

CONVERSION OF MUNICIPAL REFUSE
TO USABLE PRODUCTS BY PYROLYSIS

Figure 7.3. Destructive distillation. Organic fractions of municipal wastes can be pyrolyzed thermally to yield valuable by-products. (From Kenahan: Solid wastes. Environmental Science and Technology 7:597 (July) 1971. Courtesy of American Chemical Society.)

The community near a rendering plant may judge that the disadvantages outweigh the advantages, but the decision is based only on local considerations—the other communities whose animal wastes are being reprocessed do not participate in the decision. Thus, the trade-off rules are not laid down by ecologists; they are determined by law and by custom. We cannot predict whether or not the ecologically sound trade-off rules will begin to receive preference in time to prevent disruptions from which we cannot recover.

PROBLEMS

1. Classify the fate of each of the following manufactured products as to whether it will undergo biological recycling, industrial recycling, or accumulation as a solid waste: paste made from

casein (a milk protein); a polyethylene squeeze bottle used as a container for mustard; a copper drainpipe from a wrecked house that is sold to a dealer in scrap metals; a woolen sweater; the gold filling in an extracted tooth; the steel in an automobile body that is returned as scrap to the mill; a "no deposit" soda bottle; the coffee grounds you used to make this morning's coffee.

2. It has been suggested that we dispose of solid wastes by putting them into orbit around the Earth. What do you think of the feasibility of this plan? Defend your answer.

3. Explain the difference between an open dump and a sanitary landfill.

4. Discuss the economic and ecological trade-offs between the use of low and high land as waste disposal sites.

5. Discuss the economic and ecological trade-offs between incineration and recycling as methods of waste disposal.

The following problem requires arithmetic computation.

6. Calculate the per cent of recycling of deposit bottles when the average number of round trips per bottle was 30.

Answer

6. 30/31, or 97%

BIBLIOGRAPHY

There is considerably less literature on the topic of solid wastes than there is on air or water pollution. Most of the popular books on the environment that are cited in the bibliography of Chapter 1 include some discussion of solid wastes. For more technical information, refer to the following:

Richard B. Engdahl. *Solid Waste Processing.* Public Health Service Publication No. 1856. Washington, D.C.: U. S. Government Printing Office, 1969. 72 pp.
R. A. Fox. *Incineration of Solid Wastes.* New York: Robert A. Fox, 1967. 81 pp.
C. H. Lipsett. *Industrial Wastes and Salvage.* New York: Atlas Publishing Co., 1967.
R. D. Ross, ed. *Industrial Waste Disposal.* New York: Litton Educational Publisher, Van Nostrand—Reinhold Books, 1968. 340 pp.
Automobile Disposal—A National Problem. United States Bureau of Mines. Washington, D.C.: U.S. Government Printing Office, 1967, 869 pp.

8

THE GROWTH OF HUMAN POPULATIONS

8.1   INTRODUCTION

The 1970's promise to be a decade of unprecedented population growth. Not only will each year bring an increase in the total number of people on Earth, and not only is the rate of growth expected to become higher each year, but even the rate of the rate, the acceleration, will probably continue its current rapid ascent. Our forefathers did not live in times of such rapid change. Estimates of world populations before the 20th century are very approximate. (Indeed, even current population data are sketchy for much of Africa, for all of China, and for various other areas.) Anthropological evidence suggests that modern man evolved one hundred thousand years ago. During prehistoric times, the total human population of the Earth must have fluctuated widely. In some years there were more deaths than births, causing human population to decrease temporarily. By the first century A.D., however, population had already established its present pattern of continued growth. Rates of growth were very slow and extremely variable but each decade saw more men on Earth. The Renaissance in Europe marked the beginning of a rapid rise in world population. At the time of the discovery of America, there were about one-quarter billion people alive on Earth. In 1650, about a century and a half later, world population had doubled to one-half billion. In another 300 years, world population multiplied fivefold to 2.5 billion persons.

"Excuse me, sir. I am prepared to make you a rather attractive offer for your square." (Drawing by Weber; © 1971 The New Yorker Magazine, Inc.)

During the 1950's, the population increased almost another one-half billion and, by 1970, world population was approximately 3.5 billion persons. In other words, the *increase* in world population from 1950 to 1970 was about twice the *size* of world population in 1650. Or consider another comparison. Today there are more people in China than there were people on Earth in 1650. Or yet another: two-thirds of all people born since 1500 are alive today.

Figure 8.1 presents a schematic graph of world population size

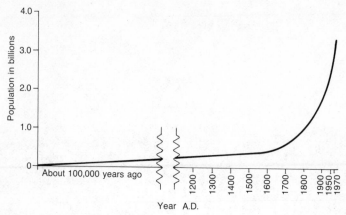

Figure 8.1. World population size from emergence of *Homo sapiens* to 1970.

since the emergence of modern man. Remember that the smoothness of the curve reflects ignorance of details of population data rather than regularity of population increase. It is clear that the curve of growth is becoming steeper and steeper in time. In fact, a glance at Figure 8.1 may well cause you to panic. If the population continues to grow ever more steeply, or even if it continues to grow at its current rate, very soon there will be too many people for Earth to support. If there is poverty and starvation now, how can economic and agricultural development be expected to keep pace with an exploding population? Destruction of land, depletion of natural resources, production of waste, and pollution of the Earth can all be expected to increase with increased population. You have probably read dire predictions based on extrapolation of the growth curve of Figure 8.1. One projection which has gained a certain currency in both lay magazines and scientific journals says that if the present rate of world population increase were to continue, there would be one person for every square foot of the Earth's surface in less than 700 years. If we accept the premise, the conclusion is necessary. But we know that the conclusion must be false. It is impossible for men to stand elbow to elbow on this planet. One square foot of Earth's surface cannot feed, clothe, and shelter a person. Thus the premise must be untenable. In other words, human population cannot continue to grow at current rates indefinitely. Indeed, similar reasoning should convince you that *population size cannot grow forever*, even very slowly. It will be checked by such factors as limits of space and food, by explicit decisions of families and nations, by famine and diseases, and by complicated interrelated social forces.

Clearly, though, it is of great importance to know where the curve in Figure 8.1 is going. One reason we need such estimates is so that we can plan for the future. How much food must be produced during the next decades? How many schools should be built? Where should roads be placed? Parks? Power plants? Without some methods for projecting future population growth, planners would have insurmountable difficulties. Lest these questions appear to imply that population growth acts as an inexorable and independent force, we must emphasize that social and economic factors help determine population size just as population size is one determinant of social and economic situations. For example, a well-educated and well-fed society typically tends to grow slowly, for people are both aware of population control and motivated to practice it. Conversely, a society that grows slowly is likely to be able to be well-fed.

A second reason that we need to know how to predict population growth is political. If we can arrive at a reasonably accurate estimate of population size at some future date, and we can show it to be too large to be consistent with societal well being, we have a numerical weapon with which to fight for the implementation of population control measures. In order to evaluate for yourself the validity of statements, predictions, and proposals concerning population size, you, too, must understand the mechanisms of population growth.

We end this introduction with four fallacious statements which serve as examples of uncritical thoughts on population. The first statement is from a recommendation of a group advocating population control; the remaining three paraphrase statements from texts on human ecology. Later, at the end of Section 8.3, we will use the tools developed in this chapter to analyze the four fallacies.

(a) To maintain a constant human population, no family should have more than two children.

(b) Modern medicine, by prolonging the lives of elderly persons, has contributed greatly to the population explosion.

(c) Since the population explosion is of worldwide concern, it is appropriate to study the growth of the world's population rather than to focus on that of a particular region.

(d) Since there is no migration to and from Earth, to study human population growth we need consider only births and deaths, and we may ignore migrations from one country to another.

8.2   EXTRAPOLATION OF POPULATION GROWTH CURVES

We noted in the previous section that the most obvious method for predicting population growth is to construct a graph that plots population size against time and to guess how the curve will continue. Guessing points on a curve outside the range of observation is called **extrapolation.** Extrapolation is a subtle art. Look at Figure 8.1. If you didn't know the labels of the axes and were asked to continue the curve, what would you do? Such an exercise is frivolous. One person might think the curve will continue to go up indefinitely; another might try to draw a curve that peaks and then falls below zero. Still another might finish it by continuing up for a while and then leveling off. And someone else might think that the curve will become wiggly and erratic. If, however, you knew you were graphing population, you could quickly exclude some kinds of curves. Infinite and negative populations would be impossible, a zero population highly improbable for many years, and wide fluctuations unlikely. Exclusions, however, wouldn't construct your curve. On the other hand, if you had a theory of population growth you would have a basis for extrapolation.

A model for a mechanism of population growth was introduced in 1798 by the Reverend Thomas Robert Malthus. He noted that population, when unchecked, grew at a **geometric rate** of increase. For example, if there were x people in year 0 and ax in year 1, (where a is greater than 1) there would be $a(ax)$, or a^2x, in year 2, a^3x in year 3, and a^nx in year n. However, Malthus said that food supplies increase at an **arithmetic rate**: that is, if there were y pounds of food in year 0, there would be $y + a$ in year 1, $y + 2a$ in year 2, and $y + na$ in year n. Table 8.1 presents examples of both types of growth. In Figure 8.2A, both types of growth curve are depicted graphically. Geometric growth is much faster than arithmetic; therefore, predicted Malthus, any population, if uncontrolled, will eventually grow too large for its food supply. He continued:

> By that law of our nature which makes food necessary to the life of man, the effects of these two unequal powers must be kept equal.
>
> This implies a strong and constantly operating check on population from the difficulty of subsistence. This difficulty must fall some where and must necessarily be severely felt by a large portion of mankind. . . .
>
> . . . The race of plants, and the race of animals shrink under this great restrictive law. And the race of man cannot, by any efforts of

TABLE 8.1. COMPARISON BETWEEN GEOMETRIC
AND ARITHMETIC GROWTH

EXAMPLE:

$x = 10$ people
$y = 100$ pounds of food
$a = 2$

Year	Geometric Growth	Arithmetic Growth
0	10 people	100 pounds
1	$2 \times 10 = 20$	$(2 \times 1) + 100 = 102$
2	$2^2 \times 10 = 40$	$(2 \times 2) + 100 = 104$
3	$2^3 \times 10 = 80$	$(2 \times 3) + 100 = 106$
4	$2^4 \times 10 = 160$	$(2 \times 4) + 100 = 108$
.	. .	. .
.	. .	. .
.	. .	. .

reason, escape from it. Among plants and animals its effects are waste of seed, sickness, and premature death. Among mankind, misery and vice.*

The premises on which Malthusian theory is based have not proved true historically. First, because of rapid improvements in agricultural techniques, man's food supply has increased much faster than arithmetically; second, geometric growth curves do not adequately describe human population increases. However, the core of the Malthusian argument cannot be ignored, for there are limits to the number of persons Earth can support, and unless growth is checked rationally, something akin to the Malthusian's "misery and vice" *will* afflict mankind.

The inadequacy of geometric curves in predicting future population sizes has led to the search for other graphical representations of growth. Observing that population sizes must have limits, ecologists often describe the growth of populations in the following terms. Consider some population which is initially very small. The very fact that it is small places it in danger of extinction, because it may not be able to recover from such setbacks as epidemics, famine, or poor breeding. Even if such factors do not totally destroy the population, they will limit its growth rate, and therefore the population will increase only slowly at first. However, once the population is established, its size will rise more rapidly as long as

*T. R. Malthus, *An essay on the principle of population*. Published originally in 1798. Reprinted in G. Hardin, ed., *Population, Evolution and Birth Control* (San Francisco: W. H. Freeman and Co., 1969), pp. 7–8.

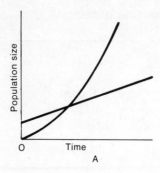

A

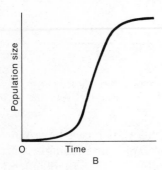

B

Figure 8.2. Schematic growth curves.
A. Arithmetic (straight line) and geometric
(curved line) patterns of growth. B. Sigmoid curve of growth. C. Growth curve of a
population that becomes extinct.

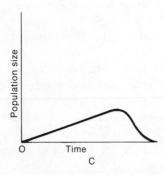

C

there is enough food, relatively few predators, and favorable living
conditions. When the population becomes large with respect both
to food supply and vulnerability to predators, and so dense that
disease spreads rapidly, the rate of growth decreases. The whole
curve of growth looks like an S and is called "**sigmoid**", or S-like.
Figure 8.2*B* shows the shape of a sigmoid curve. However, the

sigmoid curve, though often useful, does not describe all populations. Certainly populations which become extinct exhibit a very different growth curve (see Fig. 8.2C). Moreover, the sigmoid curve is rarely useful for predicting the myriad changes that occur in growth rates of human populations. For example, consider again Figure 8.1, the graph of total population size since man's arrival on earth. Even if we all agree that the curve will eventually become sigmoid, examination of Figure 8.1 gives no clue as to when the top of the S will begin to form. Indeed, if we had looked at the curve depicting growth from the beginning of man until 1600, we would have had no idea when the population would start to increase rapidly. A much more reliable method is to look, not at changes in total population size with time, but at patterns of changes in *rates of growth*. For this purpose, the reader needs to understand what rates of growth are and how birth and death rates affect the size of populations. He must be able to see the relationship between current and future growth and decline. In short, he needs to know the elements of demography.

8.3   DEMOGRAPHY

Demography is that branch of anthropology which deals with the statistical study of the characteristics of human populations, with references to total size, density, number of deaths, diseases and migrations, and so forth. The demographer attempts to construct a numerical profile of a population he is studying. He views populations as groups of people, but is not professionally concerned with what happens to any given individual. He wants to know facts concerning the size and composition of populations. For instance, he may want to know the number of males in a population or the number of infants born in a given year.

In addition to studying the composition of populations, the demographer is interested in how populations change in time. He studies changes by counting the number of **vital events** – the births, deaths, marriages, and migrations. If he knew the composition of a population at any given time and the number of vital events occurring between that time and another, he would know the composition of the population at the end of the period. For example, suppose that in 1924 the population of some village were 732. In the next two years, if there were 28 births and 15 deaths and if four people moved in and one moved out, there would be 732 + 28 − 15 + 4 − 1 = 748 people at the end of 1926.

Demographers often study **vital rates**, the number of vital events occurring to a population during a specified period of time divided by the size of the population. For example, the 1968 birth rate for the United States is the number of births in 1968 divided by the population in 1968. (Since the population itself changes during a given year, the population size is often defined to be the number of people alive at midyear, June 30.)

Before we consider demographic rates, we must understand how a population grows. If we have $100.00 in the bank accumulating 3% annual interest, we know at the end of the year there will be $103.00. We can think of the 3% operating on each dollar within the population of dollars in the bank. Therefore, each individual dollar grows to $1.03, or we can say that each dollar is growing at an annual rate of 3%. Taking one dollar out of the bank will not affect the rate of growth of the remainder.

But what does it mean to say that a population is growing at 3%? Certainly, if we know that 100 people were alive on January 1 and 103 were alive the following December 31, the annual growth rate has been

$$\frac{(103-100)}{100} \times 100\% = 3\%$$

However, if we thought of the 3% operating on each person, we would have to think of each person becoming 1.03 people. What does that mean? A 3% growth rate means only that for every 100 people there were three more births and immigrations than deaths and emigrations. Even in the very simple case in which there are only births, removal of one person from the population will affect the rate of growth of the remainder. Suppose we have 100 people on January 1, and there are three births, no deaths, and no migrations during the year. There would be 103 people on December 31, a 3% increase. Now, suppose we looked at only 99 people, omitting one infant. There would still be three births, because births do not occur to infants. So there would be $99 + 3 = 102$ persons at the end of the year. The annual rate of growth would be

$$\frac{102-99}{99} \times 100\% = 3.03\%$$

On the other hand, suppose we looked at the rate of growth of the population excluding, instead of an infant, one of the women who had a child. Now we would have two births occurring to 99 people and the rate of growth would be

$$\frac{101-99}{99} \times 100\% = 2.02\%$$

This very simple example has pointed out some of the difficulties confronting the student of population size, but it also leads to three important insights that are necessary for an effective approach to the investigation of growth:

(a) An overall "rate of growth" is really the difference between a rate of addition (by birth or immigration) and a rate of subtraction (by death or emigration). The rate of growth is positive only when there are more additions than subtractions.

(b) The probability of dying or of giving birth within any given year varies with age and sex.

(c) The age-sex composition, or **distribution**, of the population has a profound effect upon the growth rates.

To investigate population growth, it is useful to pretend that the population we are studying consists of three sub-populations defined by age. The pre-reproductive age class is defined to consist of all people under 15, the reproductive age class includes everyone from 15 to 45, and the post-reproductive class contains everyone over 45. How do vital events affect each age class? We assume that only negligibly few births occur to people outside the reproductive age class.

Death can occur to anyone. However, death rates are functions of age. We are interested, then, in age-specific death rates: that is, the number of people of a given age who die in a given year divided by the total number of people of that age. In general, these rates are high in infancy, reach their lowest values at about ten years, and then increase slowly. They become higher than infant mortality rates at some time after age 60.* Particular societal conditions or events will cause changes in this overall pattern of mortality rate. For instance, maternal mortality is very high in primitive societies. And young men in war-torn countries have abnormally

---

*Infant mortality rate is usually defined as:

$$\frac{\text{Number of deaths of infants less than one year old in a given time period}}{\text{Number of live births in the same given time period}}$$

The definition of "live birth" varies from country to country, rendering international comparisons of infant mortality rates rather tricky.

high death rates. Also, in poor countries the age-specific death rates do not reach the level of infant mortality until quite late in the life span. This age in Thailand, for example, is over 80. Figure 8.3 presents some recent mortality data for four countries. Sweden represents a nation with very low mortality throughout the life span; the United States is an example of a wealthy country with quite high infant mortality rates; Thailand exemplifies a society which lacks widely available modern medical care. The data for the African population of Guinea in 1955 shows an extremely high death rate. Note especially how the graphs of age-specific death rates cross each other. In particular, for persons over 75, the death rate for Sweden is higher than that for the United States, and both have considerably higher rates than Thailand.

It is often stated that infectious disease, famine, and natural disasters drastically limit populations. In general, this is true, but the precise effect on future growth depends upon the age of the victims of the catastrophe. If the victims are in the post-reproductive age group, the disaster will have little effect on future population size. Since often a large proportion of the victims are the elderly, allowing disasters to restrict population growth is both inefficient and immoral.

Births, though occurring to members of the reproductive age class, add new members to the pre-reproductive class. It is useful to consider only women in discussing birth rates. The **fertility** of a woman is her likelihood of producing a birth. Fertility is influenced by both biological and sociological factors. A woman is said to be **fecund** if she is biologically able to conceive and to bear children. The fecundity of a woman is zero before puberty, increases throughout her teens and early twenties, and then drops, at first slowly, then quickly, finally reaching zero. But of course women do not have babies throughout their fecund period. Certain social patterns greatly influence probabilities of giving birth. First, consider the age at which women marry. Among those primitive societies in which the average age at marriage is very young, women tend to have babies throughout their reproductive span. In societies in which marriage is postponed, the effective number of years during which women conceive is greatly reduced. In Ireland, for instance, where the median age at marriage for women who ever marry has been nearly 30 during much of this century, reproduction has been postponed until after the most fecund period. (However, in the past decade, the median age has dropped below 25.)

Other social factors that greatly affect the levels of fertility are

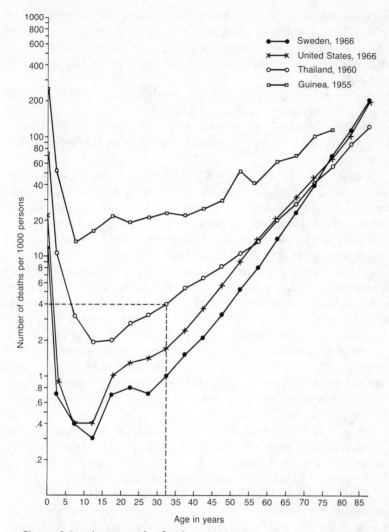

Figure 8.3. Age-specific death rates for four selected countries.

Example. The point shown at the intersection of the dashed lines tells us that, in Thailand in 1960, the death rate for people in the 30- to 35-year-old age group was four out of every 1000. This statement can also be expressed by the following equation:

$$
\text{Death rate in Thailand in 1960} = \frac{\text{Number of people in Thailand in the 30- to 35-year-old age group who died in 1960}}{\text{Total number of people in Thailand in 1960 in the 30- to 35-year-old age group}} \times 1000
$$

$$= 4$$

religious and political attitudes encouraging or discouraging births, the role of the woman in society, the amount of time spent by the father away from home, and attitudes concerning the duty of the child as producer in childhood and as supporter of his parents in their old age.

The number of births is determined not only by fertility patterns, but by the number of women in the reproductive age classes; this, in turn, depends upon the number reaching 15, on patterns of migration, and, of course, on mortality during the reproductive years.

Patterns of migration do not greatly affect growth except when a large proportion of the population migrates, especially when the migrants are predominantly of one sex. For example, in the last century many Irish young men emigrated, leaving a surplus of women behind. Consequently there were many childless women. Conversely, at the beginning of this century, there were many more men than women in the United States. The differential can be attributed in large part to the fact that the heavy immigration of the period was male-dominated. War and military service are examples of two other social patterns which temporarily or permanently remove reproductive-age men from the population.

The actual number of babies born every year depends on both the proportion of women of reproductive age producing a birth in a given year and the size of the reproductive age category. Therefore, both must be known in order to predict accurately the population changes that will take place.

In addition to considering vital events, our tripartite division of a population compels us to recall that at any given time members of a population are aging. Each year people who become 15 or who become 45 leave one age group and enter another.

Now consider a hypothetical population with fixed age-specific vital rates: for example, a population represented by one of the curves in Figure 8.3. Further, pretend that we may double the number who live through any one year of age we choose. (This assumes a population with death rates no lower than 50%. See Table 8.2 for a demonstration that doubling survivors is not the same as halving the death rate.) What age should we choose to double if we desire the greatest increase in population? Clearly, we will have most effect if we reduce infant mortality (deaths from age 0 to age 1), for if in the long run there were no other changes in rates, two times as many viable infants would lead in just a century to a doubling of population size. On the other hand, if we were to double the number of surviving 80-year-olds, the population

TABLE 8.2. WHY IS HALVING THE DEATH RATE NOT THE SAME AS DOUBLING THE SURVIVAL RATE?

EXAMPLE 1.° HALVING THE DEATH RATE.

(1) Death Rate/1000	(2) Survival Rate/1000
1000	0
500	500
250	750
125	875
·	
·	
·	
8	992
4	996
2	998
1	999

Successive halving ⟶ Decreasing change

EXAMPLE 2. DOUBLING THE SURVIVAL RATE.

Survival Rate/1000	Death Rate/1000
1	999
2	998
4	996
8	992
·	
·	
·	
100	900
200	800
400	600
800	200

Successive doubling ⟶ Increasing change

°In both examples, column (2) = 1000 − column (1)

size would hardly change. Why? First, few people reach 80; second, even healthy 80-year-olds cannot survive very much longer; finally, the aged do not produce babies.

To gain insights into the numerical effects of reduction of infant mortality, we shall compare Chile in 1934 to Sweden in 1970. In Chile in 1934 the infant mortality rate was extremely high — 262 deaths for every 1000 live births! By contrast, the infant mortality rate in Sweden in 1970 was about 13 deaths per 1000 live births. What would have happened to the population of Chile after 1934 if all birth rates and death rates had remained constant but the infant mortality had suddenly dropped to levels approaching Sweden's? It is easy to see that the sudden reduction in infant mortality with no concomitant change in birth rate would have quickly led to a rapid increase in total population. In 1934, the probability that any infant who was born in Chile would reach the age of 1 was about 75%. If he reached 1 year, the probability that he would reach 15 years was about 90%. Thus, for every 1000 babies born only about 650 reached reproductive age. On the other hand, if infant mortality had been reduced to about 13 per thousand per year, and if all other mortality rates had remained unchanged, about 900 of every 1000 infants born would have reached 15 years of age. This would have been about a 30% increase in the number of people who would have reached the reproductive age. Furthermore, if mortality rates in the entire pre-reproductive age had been reduced to the level of Sweden's, 950 out of every 1000 people would have reached 15, causing a small further increase in population growth.* Thus, lowering rates of death in the pre-reproductive years, without changing birth rates, can induce a very rapid population increase.

Historically, much of the current population explosion is due to precisely that—a rapid decrease in infant mortality without a decrease in birth rate. In Western Europe, United States, Canada, Russia, and Japan, both birth rates and death rates have decreased rather slowly over the past century. In many countries of Asia, Africa, and Latin America, introduction of modern medicine has led to changes in mortality patterns, while birth rates have remained relatively constant. Typically, medically primitive societies maintain a population balance by having both high birth and high death rates. When infant mortality drops, birth rates remain high for a while, leading to a period of time characterized by rapid growth arising from continued high birth rates and newly achieved low death rates. This period is called the time of **demographic transition**. So-called developed societies are characterized by both low birth and low death rates. They have completed the transition by dropping birth rates. The reader should understand that declines in birth rates have not usually occurred simply because of propaganda or public statements urging a change. Rather, small families have been the result of complicated social and economic forces. The developed countries have been lowering their death rates for about a century. Thus they have had many years to allow fertility patterns to change and have avoided explosive rates of growth. Furthermore, the technological and economic develop-

---

*In fact, the infant mortality rate in 1967 in Chile was 92 infant deaths per thousand live births; the percentage of children born alive who would reach 15 was about 85%.

ment of these countries has led to societal patterns which encourage, at least implicitly, a reduction in the number of births.

The underdeveloped countries remain today in a transition period. With the assistance of aid programs, missionaries, and others, these countries have been able to drop their death rates significantly in a few decades. Unfortunately, they have had only limited time and meager resources with which to develop the technological and social patterns which, in other countries, have preceded or coincided with a drop in birth rates. This lack of preparation is one reason why their current growth rates are generally viewed with alarm.

We are now equipped with enough demographic concepts to analyze the four fallacies of Section 8.1.

(a) *To maintain a constant human population, no family should have more than two children.*

Limiting family size to two children would produce a rapidly decreasing population. If family size were limited to two, there would be many fewer children than adults: some adults would not marry, and some families would have either no children, or one child; in addition, not everyone survives childhood.

(b) *Modern medicine, by prolonging the lives of elderly persons, has contributed greatly to the population explosion.*

This statement contains kernels of truth. Medical advances certainly have led to rapid population growth. However, advances in geriatrics have had relatively small effects on population size because the actual amount of added years of life can be quite small. (For example, even under optimal medical care, an 80-year-old is very unlikely to have many more years of life.) Furthermore, as mentioned previously, the elderly do not produce babies.

(c) *Since the population explosion is of worldwide concern, it is appropriate to study the growth of the world's population rather than to focus on that of a particular region.*

This statement displays a disregard or an ignorance of the complexity of population growth curves. Stating that the world's population is growing in a given year at 2.5% annually masks the fact that some countries are growing quite rapidly, some are growing slowly, and some may even be losing population. Even if we assume that the rate of growth for each country will remain unchanged for a decade, the annual world rate of growth may be very different from 2.5% in just 10 years. Table 8.3 shows some examples of the effect of different component rates of growth on the overall rate.

(d) *Since there is no migration to and from Earth, to study*

TABLE 8.3. What Will a 2.5% Annual Worldwide Population Growth Rate Become in 10 Years if all Growth Rates Within Countries Remain the Same?

Component Rates	World Growth Rate 10 Years Later*
All countries have a 2.5% growth rate.	2.5%
Half of the world's population has a growth rate of 2%, the other half 3%.	2.52%
Half of the world's population has a growth rate of 1%, the other half 4%.	2.70%

*Formula for growth rates:
Let p_1 = proportion of world's population having growth rate r_1.
and p_2 = proportion of world's population having growth rate r_2.

$$\text{World growth rate in 10 years} = \frac{p_1(1+r_1)^{10} + p_2(1+r_2)^{10} - p_1(1+r_1)^9 - p_2(1+r_2)^9}{p_1(1+r_1)^9 + p_2(1+r_2)^9} \times 100\%$$

human population growth we need consider only births and deaths, and we may ignore migrations from one country to another:

If migration were sporadic and random, this statement would have much validity. But, in fact, groups of migrants are usually quite different with respect to age, sex, and vital rates than the inhabitants of the country either from which they leave and to which they go. In particular, migrants are often healthy males of reproductive age.

8.4   PREDICTING FUTURE POPULATION SIZES

The preceding discussion has demonstrated that the rate of growth of a population depends, in large measure, on its age-sex distribution. Suppose, for example, we knew the population size of a certain town as of January 1, 1950. This information alone would be of little use in predicting any future population size; more helpful would be knowledge of the number of reproductive-age males and females, the number of children who will soon reach reproductive age, and the number of old people. A hypothetical age-sex distribution for our town is depicted graphically in Figure 8.4. Note that the graph shows that there are the same number

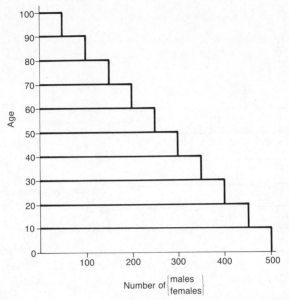

Figure 8.4. Age-sex distribution of an ideal population.

of males as females in each age group. In particular, there are 500 boys and 500 girls under 10, and 50 men and 50 women between 90 and 100 years of age. Furthermore, there are exactly 50 fewer men and 50 fewer women at each succeeding age decade. Thus, while there are 300 of each sex between the ages of 40 and 50, there are only 250 of each sex between 50 and 60.

Do human age-sex distributions look like Figure 8.4? Not at all! Figure 8.4 would represent a population in which (a) boys and girls were born with equal frequency; (b) the same number of persons were born every year for over a century; (c) everyone died by the age of 100; and (d) any person, at birth, had an equal chance of dying throughout each year of his life span. However, in real human populations, about 106 boys are born for every 100 girls.* Nor is the probability of dying constant throughout man's life span. Instead, as we have discussed in Section 8.3, a relatively large proportion of people die when they are very young, comparatively few die between the ages of 10 and 50, and the proportion of people dying each year after 50 increases rapidly. In addition, there are marked sex differences in mortality (the number of deaths which occur in a given period). Women have a higher probability of

_____

*This figure represents a worldwide average. There is considerable geographic variation.

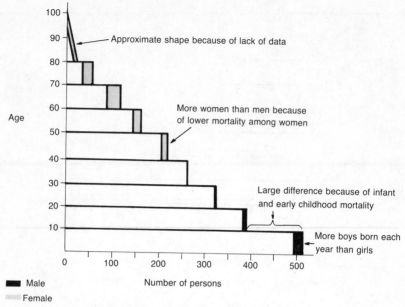

Figure 8.5. Typical age-sex distribution.

surviving from one year to the next throughout the life span except, in some societies, during the childbearing years.

Consider the effects of realistic patterns of vital events on a group of people all born within the same given year. (Such a group is called a **birth cohort**. A cohort, in general, is any group of people with something in common. A birth cohort is defined by demographers as a group of people born in a given period of time, such as a particular year.) The greater survivorship of women over men means that even though more boys are born than girls, the ratio of women to men increases as the cohort grows older. By the time the cohort is elderly, there are considerably more women than men. Also, data are usually collected in such a way that we know only the total population of each sex over 70, 80, or 85. Therefore the graphs can be only approximate for the very old age groups. Figure 8.5 presents an age-sex distribution which more nearly reflects these demographic characteristics.

In addition to these reasonably predictable phenomena, many less predictable changes can occur. Population growth is affected by such events as war, famine, medical advances, and changes in social custom. For example, Figure 8.6 presents the age-sex distribution of France in 1965. This is a very bumpy curve. To interpret it we need knowledge from many fields. For example, the

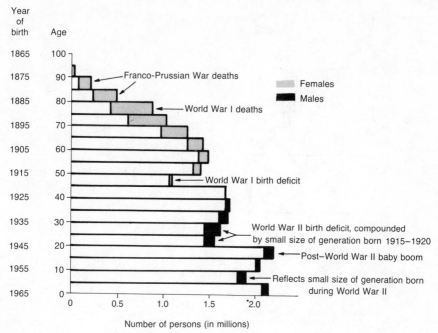

Figure 8.6. Population of France in 1965.

graph shows than in 1965 only 38% of the population of France between 70 and 74 were males. Why so few? We have already learned that men die earlier than women, but the observed discrepancy is much larger than can be explained solely by natural differences in rates. History provides an answer. This cohort was born between 1890 and 1894. At the outbreak of World War I, in 1914, the cohort born in this five-year period was between 19 and 23 years of age. Its members, then, included many of the men who fought World War I. Therefore, much of the difference in numbers of men and women in 1965 reflects the mortality of French men during World War I. We can immediately see how the distribution curves can sometimes aid in predicting future population growth. We would expect that since the males of the 1890–1894 cohort were away from home during some of their reproductive years and since so many were killed, there should occur some very small cohort about one generation later. Indeed, the birth cohort of 1915–1920, the people born for the most part during World War I, is very small.

To illustrate another example of predicting population growth from age distribution, we examine the population distribution of

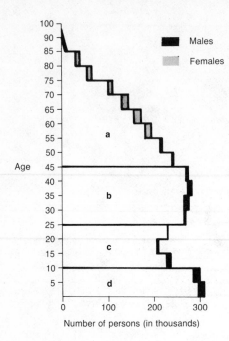

Figure 8.7. Age-sex distribution of Sweden in 1950.

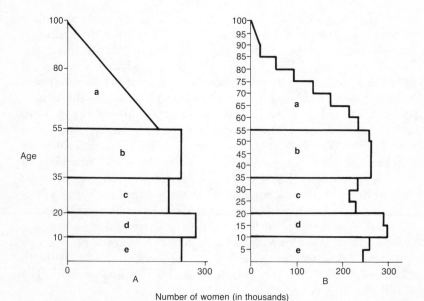

Number of women (in thousands)

Figure 8.8. Female age distribution of Sweden in 1960. *A*. Distribution guessed from age distribution of 1950. *B*. Distribution based on Swedish government population data, 1960.

Sweden in 1950 (Fig. 8.7). The total population in 1950 was nearly seven million persons. First, note that the male and female age distributions are very similar. Therefore, we need consider only one sex to examine population growth. Next, we see that the distribution is separated into four distinct parts. Since the distribution is very narrow for persons of reproductive age, we would expect the age distribution for women 10 years later to look something like Figure 8.8A. Each of the distinct sections of Figure 8.7 would move up the age axis 10 years, with each decreasing in size because of mortality, and the small reproductive class would produce a small number of births. In fact, the shape of the actual distribution curve in 1960, shown in Figure 8.8B, is remarkably similar to Figure 8.8A. The male age distribution is nearly the same shape.

A population curve need not be bumpy to be a useful predictor of growth. The age distribution for women in India in 1951, shown in Figure 8.9, looks like a triangle with a very wide base. From this graph, we would expect that such a population would be rapidly increasing. Hence, we would guess that the base of the triangle described by the age distribution 10 years later (1961) would be very wide. Figure 8.9 shows that in 1961 the base had widened considerably, that the total population had increased by about 25%, and that the proportion of the population under 10 had increased from about 25% to about 30%.* The population distribu-

---

*The deficit of 15- to 20-year-olds could not have been predicted from examining the 1951 distribution alone.

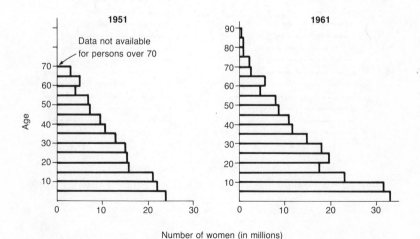

Number of women (in millions)

Figure 8.9. Female age distributions of India in 1951 and 1961.

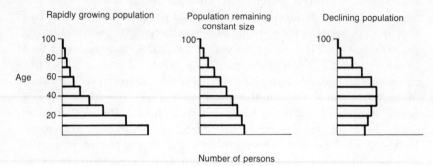

Figure 8.10. Schematic population-age distributions.

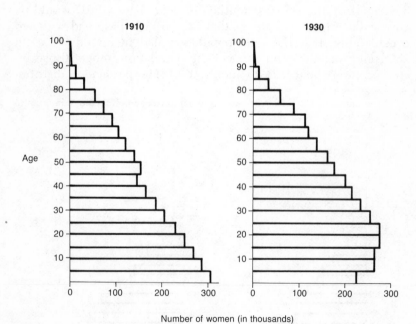

Number of women (in thousands)

Figure 8.11. Female age distributions of Sweden in 1910 and 1930.

tion of 1961 predicts an even more rapidly increasing population during the next decades.

The shapes of smooth population curves tell much about the growth of a population. Rapidly growing populations have very large bases, populations that are changing in size have relatively narrow bases, and declining populations have pinched bases. (See Figure 8.10).

Predictions about future population growth from the age of the distribution alone are not sufficient if there are drastic changes in mortality and fertility. For example, Figure 8.11 shows that the distribution for Sweden looked like a triangle in 1910. Not knowing anything about changes in vital rates, we would predict a triangle in 1930. However, in 1930 the base of the age distribution was pinched. Reexamination of Figure 8.5 shows that the pinched base persisted through the birth cohort of 1940. A demographer would have had to predict World War I and its profound economic and social effects on all of Europe to have projected accurately the population of Sweden from 1910 to 1930.

8.5   THE ULTIMATE HUMAN POPULATION

These preceding examples have shown a simple way to predict future population growth directly from current age distributions. We have also demonstrated that predictions depend on guesses about future vital rates. Demographers have sophisticated methods for guessing at changes in rates, and their predictions for short periods are often very accurate. For the more distant future however, no estimate can be made with certainty.

We cannot know what ultimately will limit the growth of human population. Pessimists predict mass starvation, total war, irremediable destruction of drinking water, a lethal upset in the oxygen–carbon dioxide balance. They cite evidence of cannibalism in overcrowded rats, suicide in lemmings. Optimists speak of rational self-limitation. However, it is reasonable to guess that limits, either cataclysmic or rational, will not occur on a worldwide basis but rather will depend on factors within smaller geographic units. Countries or areas with decreasing birth rates may achieve self-limitation with little disruption of life and environment. In countries with rapid rates of growth, where food shortages threaten, the mechanisms that will limit population will not be pleasant ones. Various setbacks are likely to precede significant decreases in birth rate and perhaps thereby avert mass death from actual starvation. For example, there may gradually occur a general state of undernutrition, leading both to increased susceptibility to disease and to lower fertility. Destruction of drinking waters may lead to widespread epidemics of contagious diseases. Technological development, if it causes increased air pollution, will lower the general health of a population. Moreover, these factors affect other forms of life as well. For example, if pure water becomes scarce, populations of many species decrease, lowering man's food supply.

Predictions about man's potential growth must take into account the interrelationship of man and his environment. When we predict the amount of growth in some small period of time, it is reasonable to ignore the populations of other species. However, when we predict far into the future, we must account for other living things. We must not think of world population growing wildly until the time when there is not enough food, air, water, or room in which to stand. Rather we would expect increasing population size to cause a gradual deterioration of levels of health and nutrition, to lead to increases in the number of areas of the world beset by poverty and overcrowding. This view is essentially Malthusian without specifying particular rates of growth. The only optimistic note in this profoundly distressing picture of the future is that the very symptoms which we have just listed will increase mortality levels and decrease birth rates and thus tend to reduce population. The question we must face is whether reduction will be fast enough. Some species, after all, become extinct because they and their environment are unable to adjust to each other. This book cannot predict man's fate. We ask the reader to recognize the seriousness of the questions and to realize that it cannot be answered by looking at graphs and blindly drawing lines.

PROBLEMS

1. Define the following terms: rate; rate of growth; rate of change.

2. Explain why world population size cannot grow forever.

3. Discuss some uses of predictions of population size.

4. Outline Malthus' theory of population growth. What is the distinction between arithmetic and geometric growth? Does the growth of money in a bank proceed by arithmetic or geometric progression? What about growth in the age of a person?

5. What is a sigmoid curve? Discuss its relevance as a model for human population growth.

6. Define demography; vital event; vital rate.

7. Discuss possible explanations for the fact that the graphs of mortality rates in Figure 8.3 cross each other.

8. Contrast fertility with fecundity. Discuss factors influencing each.

9. How does migration affect population growth?

10. How are age-sex distribution curves useful in predicting future growth? What are some of their limitations?

11. How would you expect each of the following to affect population growth? Consider which age groups are most likely to be affected by each event and the interrelationships between the event and the population change: (a) famine; (b) war; (c) lowering of marital age; (d) development of an effective method of birth control; (e) outbreak of a cholera epidemic; (f) severe and chronic air pollution; (g) lowering of infant mortality; (h) institution of a social security system; (i) economic depression; (j) economic boom; (k) institution of child labor laws; (l) expansion of employment opportunities for women.

The following questions require some arithmetic reasoning or computation.

12. Justify the formula at the bottom of Table 8.2.

13. This problem investigates family size as it relates to population growth: Consider a population where 10% of all women are childless. Roughly 10% have only one child. Assume that all others have exactly two children and that there is no migration.

    (a) How many children will a cohort of one thousand women produce?

    (b) Suppose that 90% of babies reach reproductive age. How many children reaching reproductive age will the cohort produce?

    (c) How many children who reach reproductive age must a cohort of 1000 women produce in order to exactly replace itself? (Assume equal numbers of boy and girl babies.)

    (d) How many children must be born to the cohort in order for it to replace itself, that is, to produce an eventual zero population growth?

    (e) What must be the average number of babies produced per woman in order to achieve an eventual zero population growth?

    (f) What must be the average number of babies produced by the 80% of women who have more than one baby in order to achieve an eventual zero population growth?

    (g) If women were to produce the average number of babies computed in part (e), why would a zero population growth not necessarily be achieved immediately?

14. In this problem we shall study vital rates and changes in population distribution. Let us consider a population on January 1, 1970 with the following age distribution and vital rates. Assume there is no migration.

POPULATION DISTRIBUTION: JAN. 1, 1970

AGE	POPULATION, BOTH SEXES, IN MILLIONS	DEATH RATE/1000	BIRTH RATE/1000
<1	4	150	0
1–14	40	15	0
15–44	44	10	100
45–64	10	20	0
65+	2	100	0

(a) What is the total population represented by the table?

(b) What is the overall (crude) birth rate per thousand per year?

(c) What is the crude death rate?

(d) What is the rate of natural increase, i.e., the difference in per cent, between crude birth and crude death rate?

(e) Suppose there are 3,000,000 14-year-olds, 1,000,000 44-year-olds, and 500,000 64-year-olds. Assume the death rates are the same for each age class. (For example, the mortality rate for 20-year-olds and for 44-year-olds is 10 deaths per thousand per year.) Finally, assume that all births occur on January 1 and all deaths occur on December 31. Show that the age distribution on January 1, 1971 is:

POPULATION DISTRIBUTION: JAN. 1, 1971

AGE	POPULATION, BOTH SEXES IN MILLIONS
<1	4.400
1–14	39.845
15–44	45.525
45–64	10.300
65+	2.290

(f) What is the total population on January 1, 1971?

(g) What has been the rate of growth during 1970? How is this answer related to your answer to part (d)? Explain.

(h) How has the population distribution curve changed shape during 1970? Discuss the implications of the decline in 1- to 14-year-olds. Which age group had the largest per cent of increase?

(i) In order to compute a population distribution for 1971, we made many simplifying assumptions in part (e). What were they? What would be the effects on your conclusions had more realistic assumptions been made?

(j) If there had been migration, how would the overall rate of population growth compare with the rate of natural increase, computed in part (d)?

Answers

13. (a) $[(0 \times 0.10) + (1 \times 0.10) + (2 \times 0.80)] \times 1000 = 1700$
    (b) 1530
    (c) 2000
    (d) $(2000/0.9) = 2222$
    (e) 2.2
    (f) 2.65
    (g) Depends on the current age-sex distribution. (See Problem 14)

14. (a) 100 million persons
    (b) 44
    (c) $(.04 \times 150) + (.4 \times 15) + \ldots + (.02 \times 100) = 20.4$
    (d) 2.36%
    (f) 102,360,000
    (g) 2.36%

BIBLIOGRAPHY

This chapter has introduced demographic techniques for analyzing population growth. There are several valuable texts available for those interested in further study of demography. A short, but relatively sophisticated text, is

George W. Barclay. *Techniques of Population Analysis.* New York: John Wiley and Sons, 1958. 311 pp.

A much more advanced and technical, although non-mathematical, reference, is

Mortimer Spiegelman. *Introduction to Demography,* rev. ed. Cambridge: Harvard University Press, 1968. 514 pp.

Spiegelman includes an extremely large but unannotated bibliography covering a wide range of topics related to population size, control, measurement, and so forth. Finally, for the mathematically inclined, an elegant approach to demographic analysis can be found in

Nathan Keyfitz. *Introduction to the Mathematics of Population.* Reading, Mass.: Addison-Wesley Publishing Co., 1968. 450 pp.

This highly technical and mathematical text is especially careful in its presentation of interrelationships among various measures of population composition and vital rates.

Sociological factors, as we have noted, are of crucial importance to the study of population growth. A useful introductory text which combines sociology and demography is

William Petersen. *Population*, 2nd ed. New York: Macmillan Co., 1969. 735 pp.

Petersen includes a fine annotated bibliography at the end of each chapter. A more advanced sociological discussion is presented by

James M. Beshers. *Population Processes in Social Systems*. New York: Macmillan Co., The Free Press, 1967. 207 pp.

This book is useful for learning the interrelationships between social systems and patterns of demographic transition, fertility, migration, and mortality.

Several volumes of collected papers afford most interesting reading in many areas of importance to the student of population. A fascinating collection of essays is found in

Stuart Mudd, ed. *The Population Crisis and the Use of World Resources*. The Hague: Dr. W. Junk, Publishers, 1964. 562 pp.

Another recommended reader is

Garrett Hardin, ed. *Population, Evolution, and Birth Control*, 2nd ed. San Francisco: W. H. Freeman and Co., 1969. 386 pp.

Several books sound a tocsin for our crowded planet. One of the most popular of these is

Paul R. Ehrlich. *The Population Bomb*. New York: Ballantine Books, 1968. 201 pp.

Ehrlich includes a bibliography of similar discussions.

On the other hand, there is an important argument for encouraging moderate population growth expressed in a very provocative work:

Alfred Sauvy. *General Theory of Population*. Translated by Christophe Compos. New York: Basic Books, 1969. 550 pp.

For the student interested in data sources, two works are highly recommended. The most useful and complete source of world population data is the *United Nations Demographic Yearbook*, published annually since 1948. For many nations and areas of the world the *Yearbook* includes the most recent available information on population sizes, vital rates, and many more specialized demographic statistics.

Another interesting source of data is

Nathan Keyfitz and Wilhelm Flieger. *World Population: An Analysis of Vital Data*. Chicago: University of Chicago Press, 1968. 672 pp.

The authors summarize data for several countries, extrapolate various demographic measures into the future, and perform many calculations useful to the student of population.

9

THERMAL POLLUTION

9.1   DEFINITIONS AND INTRODUCTION

For many years, the word pollute has meant "to impair the purity of," either morally* or physically.** The terms air pollution and water pollution refer to the impairment of the normal compositions of air and water by the addition of foreign matter such as, for example, sulfuric acid. Within the past few years two new expressions, thermal pollution and noise pollution, have become common. Neither of these refer to the impairment of purity by the addition of foreign matter. Thermal pollution is the impairment of the quality of environmental air or water by raising its temperature. The relative intensity of thermal pollution cannot be assessed with a thermometer, because what is pleasantly warm water for a man can be death to a trout. Thermal pollution must therefore be appraised by observing the effect on an ecosystem of a rise in temperature. Similarly, noise pollution has nothing to do with purity: foul air can be quiet, and pure air can be noisy. Noise pollution (to be discussed in the next chapter) is the impairment of the environmental quality of air by noise. Neither of these expressions is currently in the dictionary; whether or not they become accepted for usage depends on how widely they continue to be employed in speech and writing.

---

*(1857.) Buckle, *Civilization*, I., viii, p. 526: "The clergy . . . urging him to exterminate the heretics, whose presence they thought polluted France."

**(1585.) T. Washington, trans. *Nicholay's Voyage*, IV; ii; p. 115: "No drop of the bloud should fall into the water, least the same shuld thereby be polluted."

The circumstances that are involved in the production of heat by working engines lead us to some rather subtle concepts in the branch of physics called **thermodynamics** (*thermo* referring to heat and *dynamics* to work). It is important to explore these concepts if we are to appreciate the nature of thermal pollution. Also, it has been said that no person can be considered to be liberally educated if he does not understand the Second Law of Thermodynamics. The first part of this chapter, therefore, should help the reader to understand why thermal pollution is inevitable and should remove at least one obstacle to his educational fulfillment.

9.2   TWO QUESTIONS ABOUT HEAT

Oil is fed to a burner, where it reacts with oxygen from the air to produce heat.

Oil is fed to a diesel engine, where it reacts with oxygen to do work, such as moving a truck up a hilly road. But it also produces heat—the working engine is always warmer than its surroundings.

Oil is fed to a power plant, where it reacts with oxygen to do work, such as driving a generator to produce electricity. But, again, it also produces heat.

In this chapter we consider two questions: (a) Is it inevitable that heat be produced, whether we want it or not, when work is done? (b) How does the production of heat affect the environment? These questions are by no means simple; they lead us to rather subtle and far-reaching concepts. To answer the first question, we will start by examining the nature of work.

9.3   WORK BY MAN, BEAST, AND MACHINE

We are going to describe an imaginary situation—a problem— that can be solved only by doing work. We will then consider what methods men might use to do the necessary work and how these methods affect the production of heat.

The problem is this: A landslide causes a large boulder to roll downhill; it comes to rest in a position where it blocks the entrance to a cave, and a man wishes to remove the obstruction. What can he do? Primitive man might have tried to push it, roll it, or drag it. If he lacked the strength to get it out of the way, he would have

been forced to search for a new cave. However, if he could cooper-
ate with his fellows, perhaps a group of men, working together,
could have been able to push or roll or drag it out of the way. The
situation would be still further improved after man had learned to
domesticate large animals and use them as beasts of burden and
of work. But a much more far-reaching development for doing work
would be the invention of machines. The lever, the wheel, the
roller, the screw, the inclined plane, and, later, the gear and the
block and tackle were all devices that could increase the force
that a man or his animals could exert.

These three techniques — cooperation among people, the use
of animal power, and the application of machines — enabled man
to build (and sometimes to destroy) such wonders as the pyramids
of Egypt, the Great Wall of China, and the cities of Athens, Rome,
and Carthage.

We will now discuss other ways to get work done. But do not
forget the cave man and his boulder; we will come back to him.

9.4 THE FIRST LAW OF THERMODYNAMICS (OR "YOU CAN'T WIN")

As machines became more and more effective in extending the
force of living muscle, inventors saw no reason to doubt that,
by providing machines that would produce work indefinitely just
because their mechanisms were so clever, continued improvement
in design would eventually free man and his animals entirely from
their labor. If our cave man had such a device (which we call a
"perpetual motion machine of the first kind"), he could move all
the boulders he wished at his leisure. It may be difficult for the
modern reader to appreciate the fact that this objective seemed
entirely reasonable, apparently requiring only continued progress
along the lines that had already been so successful.* However,
all attempts failed. The failures have been so consistent that
we are now convinced that the effort is hopeless. This conviction
has been stated as a law of nature and is called the **First Law of
Thermodynamics.**

This law can also be expressed in terms of conservation of
energy by the statement, "energy cannot be created or destroyed."

Don't ask for a proof of the First Law. There is none. The First
Law is simply a concise statement about man's experience with

---

*Before you laugh at what you may think was the naiveté of the attempts at
"perpetual motion," try problem 4 at the end of this chapter.

machines. If it is impossible to create energy, then it is hopeless to try to invent a perpetual motion machine, and we may as well turn to some other method of doing work.

9.5   HEAT ENGINES AND THE SECOND LAW OF THERMODYNAMICS (OR "YOU CAN'T BREAK EVEN")

Modern technology demands far more energy than can be supplied by men and beasts, even with the help of machines. Instead, it makes use of heat engines, which consume fuel to produce heat, and convert the heat into work.* The idea that heat could be converted into work was far from obvious. In fact, heat engines have been used successfully only during the past 200 years or so. (James Watt developed his steam engine in 1769.) The first experimental proof that energy can be converted, without gain or loss, from one form to another was supplied by James Joule in 1849. Heat and work are two forms of energy, and it is therefore possible to convert heat into work or work into heat. Thus one can create heat by rubbing two sticks together, and the heat produced is exactly equivalent in energy to the work required to rub the sticks. It was also discovered that fuels contain potential heat. Thus, a pound of coal contains stored energy which can be released by combustion. But heat is also stored in substances that are not fuels—in such ordinary substances as water. Water loses energy when it turns to ice; therefore, water must *have* energy. Why not, then, use this energy to drive a machine to do work? Such a machine, although seemingly not quite so miraculous as the perpetual motion machine of the first kind, would extract energy from its surroundings (for example from the air or from the ocean) and convert it into useful work. The air or water would then be cooled by the extraction of energy from it, and could be returned to the environment. Automobiles could then run on air, and the exhaust would be cool air. A power plant located on a river would cool the river while it lighted the city. Such a machine would *not* violate the First Law, because energy would be conserved. The work would come from the energy extracted

---

*Of course, a man (or an ox) is also a heat engine. He consumes food, which is his fuel, and converts its energy into the work of muscle contraction. Mechanical heat engines, however, can be much larger than animals, and can consume fuels such as gas, oil and coal that animals do not eat. The result is that the total amount of work done and heat produced is increased to an extent that has new effects on the environment.

from the air or water, not from an impossibly profitable creation of energy. Such a device is called a "perpetual motion machine of the second kind;" alas, it too has never been made and we are convinced that it never will be.

Let us return now from the impossible to heat engines that use fuel, where the situation continues to be discouraging. It was learned through experiment that the potential energy inherent in a fuel could never be completely converted into work; some was always lost to the surroundings. We say "lost" only in the sense that this energy was no longer available to do work; what it did, instead, was to warm the environment. Ingenious men did try to invent heat engines that would convert *all* the energy of a fuel into work, but they always failed. It was found, instead, that a heat engine could be made to work *only* by the following two sets of processes: (a) Heat must be absorbed by the working parts from some hot source. The hot source is generally provided when some substance such as water or air (called the "working substance") is heated by the energy obtained from a fuel, such as wood, coal, oil, or uranium. (b) Waste heat must be rejected to an external reservoir at a lower temperature.

A heat engine cannot work any other way. The original form of this negative statement, as made by Lord Kelvin (1824–1907) is, "It is impossible by means of inanimate material agency to derive mechanical effect from any portion of matter by cooling it below the temperature of the coldest of the surrounding objects." This is an expression of the **Second Law of Thermodynamics**.

To help us gain further insight into this very fundamental concept, we return for the last time to our cave man. Imagine that he discovers he can move his boulder by wedging a bar of copper between it and the cliff and then heating the bar with a flame. Because heated metal expands, and the cliff is stationary, the boulder will move. He has constructed a basic heat engine. Now let us assume the following circumstances:

(a) The cave man has several copper bars, each 10 meters long.

(b) It is very awkward to build a fire under the copper bar between the boulder and the cliff, but it is convenient to have a fire nearby to heat the copper bars to a temperature of 100°C.

(c) The outside temperature is 20°C.*

(d) The bar expands 0.17 millimeter for every degree of temperature rise.

The cave man, thinking he has discovered that hot copper bars can do work, heats one of his bars on the fire, wedges the hot bar

---

*Refer to the Appendix for a discussion of temperature.

between the cliff and the rock, then waits and watches. The bar cools and contracts, and no work is done. This failure teaches him his first lesson in the design of heat engines: Work is done only by heating materials and allowing them to expand. Therefore, temperature differences rather than absolute temperatures are important. Having learned this, our cave man now decides to wedge a *cold* bar between the rock and the cliff and to bring hot bars to the cold one. Let us say that he heats three bars to 100°C (we will ignore heat losses to the air). He places the first hot bar on the cold one, as shown in Figure 9.1. The cold bar gets warmer and expands, while the hot cools and contracts. When both reach the same temperature, all heat transfer (and hence all work) stops. The temperature of the working bar, which was originally 20°C, rises 40 degrees while the temperature of the heating bar drops 40 degrees. The final temperature of both bars is 60°C. In order to move the rock further, he places the second hot bar on top of the working bar and observes that the working bar warms from 60° to 80°, and the new hot bar cools from 100° to 80°. Repeating the process a third time, the working bar heats from 80° to 90°. Recall that for every degree rise, the working bar expands 0.17 mm. Thus, the first time he heats the bar the rock moves 6.8 mm, the second time it moves 3.4 mm, and the last time it moves only 1.7 mm. Therefore, the amount of work that can be extracted from a given quantity of heat depends on the temperature difference between the hot and the cold bars. This limitation is *not* imposed by the First Law; if the hot bar were

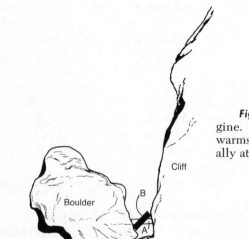

Figure 9.1. A simple heat engine. Bar *A*, originally at 20°C, warms up to 60°C. Bar *B*, originally at 100°C, cools down to 60°C.

Cliff

B

Boulder

A

to cool to 20°C and the cold one warm to 100°C, their total energies would be conserved. But such an event is contrary to all experience. In fact, the universal observation that two bodies in thermal contact eventually reach a common temperature can be taken as another expression of the Second Law.

Of course, if our cave man were clever, he would not need to tolerate a diminished output of work after the use of each successive hot bar. Instead, after each heating he would cool his working bar back to 20° in some nearby stream, replace it on the cliff (with an additional stone wedge to make it tight), and move the boulder a full 6.8 cm for every hot bar he used. Perhaps we would then think of him as a modern rather than a cave man, because his advanced technology would have caused his stream to suffer thermal pollution.

Of course, modern man uses more complicated heat engines, but the relationships between work and heat have not changed. The working substance is generally a gas that expands to do work against a piston or the blade of a turbine. The gas cools as it does its work, but it cannot cool below the temperature of its surroundings. Therefore some provision *must* be made to remove waste heat from a heat engine. These relationships are shown in a schematic way in Figure 9.2.

In practice, cooling is generally accomplished either by blowing air past the engine or by running water over it. In most automobiles, for example, water is circulated through the engine, cooled by air as it flows through the radiator, and recirculated. In large power plants, where a great deal of heat is generated, the coolant is often a river, a lake, or the ocean.

9.6   THE EFFECT OF TEMPERATURE CHANGES ON LIFE

The processes of life involve chemical reactions, and the rates of chemical reactions are very sensitive to changes in temperature. As a rough approximation, the rate of a chemical reaction doubles for every rise in temperature of 10 Celsius degrees, which is equal to 18 Fahrenheit degrees. We know that if our own body temperature rises by as much as 5 Celsius degrees (or 9 Fahrenheit degrees, which would make a body temperature of $98.6 + 9 = 107.6°F$) the fever may be fatal. What, then, happens to our system when the outside air temperature rises or falls by about 10 Celsius degrees? We adjust by internal regulatory mechanisms that main-

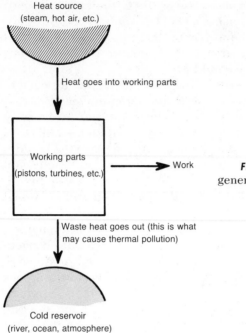

Heat source
(steam, hot air, etc.)

Heat goes into working parts

Working parts
(pistons, turbines, etc.) ⟶ Work

Waste heat goes out (this is what
may cause thermal pollution)

Cold reservoir
(river, ocean, atmosphere)

Figure 9.2. Schematic of a generalized heat engine.

tain a constant body temperature. This ability is characteristic of warm-blooded animals, such as mammals and birds. Thus the body temperature of a man or a dog in a room cannot be determined by reading the wall thermometer. In contrast, non-mammalian aquatic organisms such as fish are cold-blooded: that is, they are unable to regulate their body temperatures as efficiently as warm blooded animals.

How, then, does a fish respond to temperature increase? All its body processes (its metabolism) speed up, and its need for oxygen and its rate of respiration therefore rise. Above some maximum tolerable temperature, death occurs from failure of the nervous system, the respiratory system, or essential cell processes. According to the Federal Water Pollution Control Administration, almost no species of fish common to the United States can survive in waters warmer than 93°F. The brook trout, for example, swims more rapidly and becomes generally more active as the temperature rises from 40° to 48°F. In the range from 49° to 60°F, however, activity and swimming speed decrease, with a consequent decline in the trout's ability to catch the minnows on which it feeds. This inactivity is more critical because the trout *needs* more food to maintain its higher metabolic rate in the warmer water. Outright death occurs at about 77°F. In addition, spawning and other reproductive mechanisms of fish are triggered by such temperature changes as the warming of waters in the spring. Abnormal changes, to which the fish is not adapted, can upset the reproductive cycle.

As a result of these circumstances, a rise in the temperature of a body of water can cause a replacement of one fish population with another. Examples of lethal temperatures for various species of fish in Wisconsin and Minnesota are the following: trout, 77°F; white sucker, 84–85°F.; walleye, 86°F; yellow perch, 84–88°F; fathead minnow, 93°F. One of the largest single cold-water environments in the United States, the Columbia River, is now only a few degrees below the temperature that is likely to convert its fish population from trout and salmon to walleye, smallmouth bass, and other less desirable species.

Various observations have been cited as evidence that our concern about the effects of thermal pollution is exaggerated and that heated water can, in fact, be beneficial to fish populations. For example, fish are found in waters at temperatures above their supposed lethal level. The heated water discharges from power plants, as they fan out into rivers or the ocean, are sometimes seen to be teeming with fish, as though the fish had been attracted to a beneficially warm environment. However, one must realize that an excessively high temperature is not immediately lethal; the effects are often delayed for several hours or days. Furthermore, the dense population seen in the warm power plant discharges often consists of rough, less desirable fish rather than the more valuable varieties that prefer colder waters.

In general, not only the fish, but entire aquatic ecosystems are rather sensitively affected by temperature changes. Any disruption of the food chain, for example, may upset the entire system. If a change of temperature shifts the seasonal variations in the types and abundances of lower organisms, then the fish may lack the right food at the right time. For example, immature fish (the "fry" stage) can eat only small organisms, such as immature copepods. If the development of these organisms has been advanced or retarded by a temperature change, they may be absent just at the time that the fry are totally dependent on them. It may be very difficult to predict such effects by studying stream temperatures, because a very large portion of the total flow of a given stream may be bypassed through a power plant to carry away its waste heat. While fish are easily excluded by screens from such undesirable detours, it is not easy to keep out the microscopic organisms that do make up an important part of the food chain. These organisms are subjected to temperatures that will exceed the maximum temperature of the stream. We do not yet know how serious the consequences may be—the indications of various studies range from a zero effect to a 95 per cent kill of the plankton.

As we have learned from Chapter 6, the life in an aquatic eco-system is greatly influenced by the growth of algae. The combination of thermal pollution with increased nutrient can lead to rapid and excessive algal growth, with consequent acceleration of eutrophic and other undesirable effects that we have previously discussed.

Higher temperatures often prove to be more hospitable for pathogenic organisms, and thermal pollution may therefore convert a low incidence of fish disease to a massive fish kill as the pathogens become more virulent and the fish less resistant. Such situations have long been known in the confined environments of farm and hatchery ponds, which can warm up easily because the total amount of water involved is small. As the thermal pollution load in larger bodies of water increases, so will the potential for increased loss of fish by disease.

Man-made poisons, too, become more dangerous to fish as the water temperature rises. First, all biochemical effects, including those manifested by toxicosis, are speeded up at higher temperatures. Second, warm water favors increased growth of plant varieties such as algae; and if these are to be destroyed by chemical poisons, as is often necessary with industrial or domestic water supplies, more poison is needed. Thus, in warmer water, not only are fish likely to resist poisons less, but they are also likely to be exposed to them more.

As we have already explained, the electric power industry, and particularly nuclear power plants, necessarily produce waste heat, and such heat is most conveniently discharged into flowing waters.* It is estimated that by the year 2000 the rate of cooling water needed by the power plants will be equivalent to one-third of the total rate of fresh water run-off in the United States. Since this heat could not easily be distributed uniformly among all the large and small bodies of water in the country, but would be concentrated either at shorelines or along rivers of sufficient capacity to accommodate large plants, the result may be that many United States rivers become uninhabitable for fish.

---

*Hydroelectric plants, which are essentially non-polluting, are excluded.

Figure 9.3. Effluent from a power plant on the Connecticut River. Three large pipes discharge heated water (87°F) that spreads across the river at slack tide and tends to flow downriver at ebb tide. (Courtesy of Barnes Engineering Co.)

9.7 NON-MIRACULOUS SOLUTIONS

We have learned that we cannot invent our way out of thermal pollution. As long as we generate power we will produce excess heat, and we must deal with this problem by non-miraculous means.

The waste heat from power plants must go to the environment by one pathway or another. Along the way the heat might serve some useful function, such as warming buildings, or desalinating sea water by evaporation, or providing enough warmth in large greenhouses to grow vegetables in winter. However, the cost of piping hot water is high, and no economically feasible methods for utilizing such heat have yet been worked out. The situation is inherently discouraging—waste water is hot enough to damage an aquatic ecosystem, but not hot enough to be attractive for commercial use.

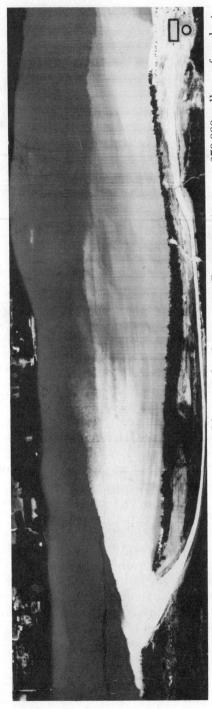

Figure 9.4. Nuclear power plant at Haddam on the Connecticut River empties up to 370,000 gallons of coolant water a minute through a discharge canal (bottom) into the river. In this aerial thermogram, made by HRB-Singer, Inc., for the U.S. Geological Survey's Water Resources Division, temperature is represented by shades of gray. The hot effluent (white) is at about 93°F; ambient river temperature (dark gray) is 77°F. (Courtesy of U.S. Geological Survey.)

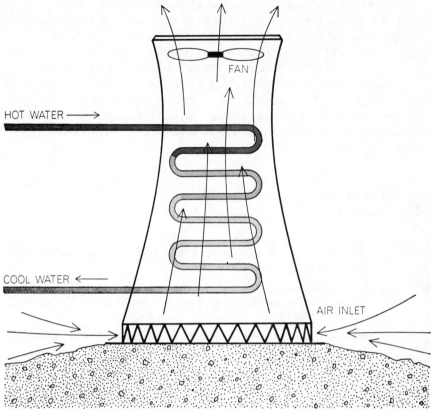

Figure 9.5. Cooling tower. Hot water flows through tubing that is exposed to air flow and is thus cooled. From Clark, J. R.: Thermal Pollution and Aquatic Life. Copyright © 1969, by Scientific American, Inc. All rights reserved.

The only other conceivable alternative is to dispose of the heat into the air.* Air has much less capacity per unit volume for absorbing heat than water does—so much less that direct air cooling of large power plants is not feasible. For this reason, such plants must still be sited near a source of water—the only other available coolant. However, the water can be made to give up some of its heat to the atmosphere before it is returned to the river or sea from which it was taken. There are various devices available that can effect such

---

*We may question whether *that* would be just another type of ecological disruption. But the heat situation in the atmosphere is different from that in water, because it is complicated by the effects of reactions of air pollutants with solar radiation. See Chapter 5 for a discussion of this problem.

a transfer, such as cooling towers or artificial cooling lakes. We will not discuss the details of these devices here, but it is important to emphasize that they all cost money to set up and to operate and that this cost would eventually appear as higher electric bills.

We are therefore left with the harsh reality of the trade-off. How much are we willing to pay for how much benefit in the form of protection of the ecosystems of Earth? And in what coin? More dollars per month? A lower setting on the thermostat for the same price? Fewer electrically powered appliances? Or the same conveniences and costs for fewer people?

PROBLEMS

1. What is thermal pollution? How does it differ in principle from air or water pollution?

2. Write as many statements as you can of the First Law of Thermodynamics. (One statement appears in Section 6.6 of the chapter on water pollution.)

3. Write as many statements as you can of the Second Law of Thermodynamics.

4. Fig. 9.6 shows a design for perpetual motion machine based on osmosis. The two identical membranes are permeable to water but not to sugar. Water from the reservoir passes up through membrane #1 to a height that is determined by the osmotic pressure of the solution. Water also permeates membrane #2, falling back to the reservoir and doing work on the way down. Do you think this machine will work? If not, why not?

5. Dogs live on all parts of the Earth from the tropics to the Arctic. Trout, on the other hand, are confined to waters no warmer than about 60°F. Why are the permissible temperatures for trout so much more limited?

6. Since marine life is abundant in warm tropical waters, why should the warming of waters in temperate zones pose any threat to the environment?

7. Outline the methods by which thermal pollution can be controlled or reduced.

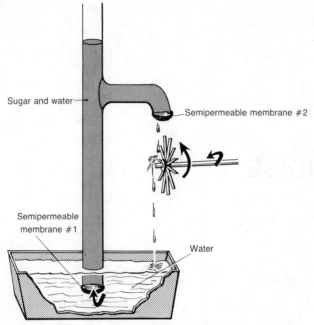

Figure 9.6.   Design for a perpetual-motion machine.

BIBLIOGRAPHY

A basic source of information on the measurement and control of temperature is the following rather massive three-volume work:

American Institute of Physics. *Temperature: Its Measurement and Control in Science and Industry.* New York: Litton Educational Publisher, Van Nostrand-Reinhold Books, Vol. 1, 1941. 1362 pp. Vol. 2, 1955. 477 pp. Vol. 3 (Parts 1, 2, and 3), 1963, 1962, 1963.

The subject of thermodynamics is covered in many standard texts at various different levels, and no specific references need be given here. "Thermal pollution" itself is a relatively new subject. Some interesting aspects that deal specifically with nuclear power plants are treated in the following book, to which we have already referred in the bibliography of Chapter 4.

Harry Foreman. *Nuclear Power and the Public.* Minneapolis: University of Minnesota Press, 1970. 272 pp.

10

█ NOISE

10.1   SOUND

To be most sensitively aware of sound, it is best to experience
silence. Think of the quietest occasions of your life. Perhaps standing in the woods after a snow on a windless winter day was one
such instance. Certainly you would not think of an occasion on
which you were moving, whether walking, running, or riding,
because motion itself produces sound. In fact, the correlation of
motion and sound is suggested by the use of the word *still,* which
means both motionless and quiet. Motion is related to energy—any
moving body has an energy of motion that depends on its speed;
the faster it moves, the more energetic it is.* Since sound is related
to motion, and motion to energy, it is reasonable to think that
sound is a form of energy; and indeed, so it is.

How is sound transmitted from a source such as a bell to a
receiver such as your ear? We note that when the bell is struck,
it vibrates. When the vibration stops, so does the sound. If the bell
is struck in a vacuum, it will also vibrate, but no sound can be de-
tected. From this experiment it can be deduced that the motion of
the bell somehow moves the air, which then moves some receiving
device in the ear. This transfer of sound energy through air occurs
in the form of a wave. We usually visualize wave motion as water
waves, especially ocean waves striking a beach, or ripples in a
pond or swimming pool. Water that is still ("quiet" water) has a
smooth, level surface. Water waves are disturbances; they alter

---

*The exact relationship is: Energy $= \frac{1}{2} \times$ mass $\times$ (velocity)^2

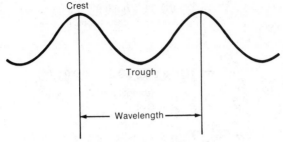

Figure 10.1. Wave length.

the normal level so as to make it higher in some places and lower in others. The highest places are called **crests**; the lowest, **troughs**. The distance between successive disturbances of the same type, such as between neighboring crests, is called the **wavelength** (Fig. 10.1). The rate at which a disturbance moves is the **speed of the wave.** The number of disturbances that pass a given point per unit time is the **frequency.** The relationship among these three attributes is

$$\text{speed} = \text{wavelength} \times \text{frequency}$$

All of these characteristics of the water wave also apply to the sound wave, except that the nature of the disturbance is different. Instead of manifesting itself as crests and troughs, as in disturbances of the water level, the sound wave is a succession of compressions and expansions that disturb the normal density of the medium (such as air) in which they are propagated. This type of wave is called an **elastic wave**, and can be illustrated by the action of a coiled spring, as shown in Figure 10.2. Imagine that a bump on a rotating wheel hits the end of the spring twice per second. The resulting compressions, and the expansions which follow them, travel along the spring at a speed that depends on the properties of the spring (not on the rate of rotation of the wheel). Let us say that this rate is 1 foot per second. The frequency must be 2 beats per second, because that is established by the speed of rotation of the wheel. Therefore the wavelength is computed as follows:

$$\text{wavelength} = \frac{\text{speed}}{\text{frequency}} = \frac{1 \text{ ft per sec}}{2 \text{ per sec}} = \frac{1}{2} \text{ ft}$$

Air is a springy substance; a squeezed balloon snaps back when released. Therefore, elastic waves can be propagated in air;

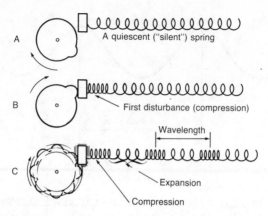

Figure 10.2. Elastic waves. A. Wheel starts to rotate twice per second. B. First impact. C. Continuous production of waves. Frequency = 2 per second. Wave speed = 1 foot per second.

this is sound. Work must be done to beat out the successive compressions; thus, sound is a form of transmission of energy. The speed of sound in air under normal conditions on earth is about 1100 ft/sec. Any object, such as an airplane, that travels slower than sound is said to be **subsonic**; one that is faster is **supersonic**.

The tone, or the pitch, of any given sound is determined by the frequency of the waves that produce it. The energy of a given sound, however, is not determined by the frequency, wavelength, or wave speed. Thus, if you slap the surface of a pond gently once per second with a spoon, you will make waves at a frequency of 1 beat per second. If you slap the water hard once per second with a paddle, you will still make waves at a frequency of 1 beat per second, and they will not travel any faster, but they will be bigger waves. The disturbances, or the heights of the crests and depths of the troughs, will be greater. The difference in height between the crest and the trough is called the **amplitude** (Fig. 10.3). The energy of a wave thus depends on the work that must be done to create the disturbance.

To think of sound energy, imagine a four-engine plane with only one engine running. The frequency of the sound depends on the type of engine and its operation; the speed of the sound is a property of the air, not of the engine. Now we start the other three engines, so that they operate just like the first. The frequency of the sound has not changed, nor has the speed of the wave. But we are doing four times as much work. (Our fuel bill will be four times as high.) The sound energy being created is therefore four

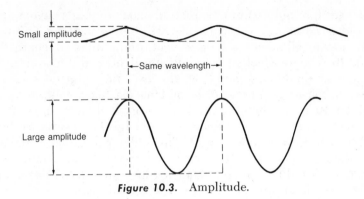

Figure 10.3. Amplitude.

times as great. Sound energy is related to loudness, but the two are not the same. We shall discuss loudness in more detail later, in Section 10.3.

10.2 NOISE

The complete physical description of a given sound cannot establish whether you, as an individual, will like it or not. If you don't like a sound, that sound is noise. In fact, noise can be defined by two words: unwanted sound. But this apparently simple concept conceals many unexpected subtleties. A given sound may be music to one person but noise to another, pleasant when soft but noise when loud, acceptable for a short time but noise when prolonged, intriguing when rhythmic but noise when randomly repeated, or reasonable when you make it but noise when someone else makes it. Of all the attributes that distinguish between wanted and unwanted sound, the one that we generally consider the most significant is loudness. There is ample evidence that exposure to loud sounds is harmful in various ways, and most people don't want to be harmed; therefore, the louder a sound is, the more likely it is to be considered noise (with some exceptions in discothèques).

10.3 LOUDNESS AND THE DECIBEL SCALE

Loudness, we have said, is related to energy. Consider a machine that converts energy into sound—for example, a siren. Imagine that you have a 50-watt electric siren. This will consume as much energy in a given time as any other 50-watt device, such as a 50-watt light bulb. Your electric bill, which requests payment for the electrical energy you consume, will be the same for a 50-watt bulb glowing all month as for a 50-watt siren wailing all month. But the bulb is more efficient at producing light, and the siren more efficient at producing sound. Of the two, the siren does its job better, because the energy per second that it actually emits as sound is about 10 watts, the other 40 being wasted as heat. In contrast, the energy per second actually emitted by the 50-watt bulb as visible light is only about 1 watt. This difference explains, in part, why light bulbs feel hotter than sirens. We conclude from these facts that it would be correct to rate loudness in watts. For example, it has been determined that a piano is as loud as an 0.1-watt siren, that a symphony orchestra is as loud as a 10-watt siren, and that a large rocket engine is as loud as a 10-million-watt siren. What is remarkable about loudness is its very wide range. A very soft whisper of a human voice has a loudness equivalent to only one-billionth of a watt. Therefore, if we were to express loudness in watts, we would suffer the inconvenience of using both very small and very large numbers. Furthermore, the scale would be awkward, for it would not start at zero, but at some small number that represents the softest audible sound. To remedy these defects and simplify the measure of loudness, we do the following:

(a) Create a new unit of loudness measurement called the decibel (*deci* meaning 10, a reference to the base of common logarithms, and *bel* after Alexander Graham Bell), and define it in terms of the ratio of the loudness of one sound to another.

(b) State that the softest sound audible to the ear will have a value of zero decibels (dB) and will be the starting point of our scale of sound measurement.

(c) Use the logarithms of the ratios of sounds of different loudnesses to express degrees of loudness. Note from the values previously given that the ratio of a very loud sound (for example, 10,000,000 watts) to a very soft sound (such as 0.000000001 watts) is a very large and inconvenient number to work with. To get smaller, more convenient numbers we shall use the logarithm of the ratio. Recall that the *logarithm* of a number "to the base 10" is simply the *number of times 10 is multiplied by itself* to give the number. As we can quickly see, when the number is a multiple of 10, its log is simply the number of zeros it contains. Thus,

10 has 1 zero, therefore log $10 = 1$
100 has 2 zeros, therefore log $100 = 2$
1,000,000 has 6 zeros, therefore log $1,000,000 = 6$
1 has no zeros, therefore log $1 = 0$

(d) Create an equation which defines loudness as follows:

Loudness in decibels of any given sound$=$

$$10 \times \log_{10} \left(\frac{\text{power}^* \text{ of the given sound}}{\text{power of a barely audible sound}} \right)$$

Example 10.1. The sound of a vacuum cleaner in a room has 10 million times the power of the faintest audible sound. What is the loudness of the sound in decibels?
Answer:

$$\left(\frac{\text{sound power of vacuum cleaner}}{\text{faintest audible sound}} \right) = 10,000,000$$

log $10,000,000 = 7$

Loudness $= 10 \times$ log $10,000,000$
$= 10 \times 7$
$= 70$ decibels

Example 10.2. What is the loudness in decibels of the faintest audible sound?
Answer: This had better turn out to be zero, which is what we promised the log scale would provide:

$$\left(\frac{\text{power of the faintest audible sound}}{\text{power of the faintest audible sound}} \right) = 1, \text{ and log } 1 = 0$$

Thus,

Loudness $= 10 \times$ log $1 = 10 \times 0 = 0$ decibels

---

*Power is energy per unit time, usually expressed in watts. The areas of the two sound sources must be the same for the comparison to be fair. The "power of a barely audible sound" is usually taken to be 10^{-12} watts/meter^2, or one-trillionth of a watt per square meter.

All this might seem quite straightforward, but there are complications. A mechanical ear, such as a microphone, converts sound to electrical power, and this power can be read out in watts, or decibels. A device that performs such a task, called a "decibel meter," is insensitive to pitch; it measures only sound power. Human beings, when asked to perform this task, generally introduce a certain subjectivity into their ratings. People generally tend to rate high-pitched sounds as being louder than low-pitched ones of the same power. Thus, a scale of loudness as judged by people would be frequency-dependent and therefore different from that of the decibel meter.

To reconcile this difference between man and machine, a subjective loudness scale has been developed. This scale is based on a comparison of any given sound with that of a pure tone having a frequency of 1000 beats (or "cycles") per second. Let us say that we wish to rate the subjective loudness of some particular sound — for example, that of a motorcycle at a distance of 25 feet. A given person is offered a selection of pure 1000-cycle sounds of different loudnesses, and is asked to pick the one that sounds just as loud as the motorcycle. Let us say that he chooses a 90-decibel sound as the best match to his ears. If this choice represents the typical, or the average response of many people, then the motorcycle at 25 feet is said to have a "loudness level" of 90 **phons**, even though its own decibel rating in terms of sound power might be some other number, such as 88 decibels. The **phon** is thus a unit of subjective loudness, which is matched to a decibel scale for a pure 1000-cycle sound.

We must, of course, consider the subjective description of loudness. If you told someone who was unfamiliar with loudness scales that a motorcycle had a loudness level of 90 phons, or produced a loudness of 88 decibels, it would mean nothing to him. It would be more informative, if less precise, to say that a motorcycle sounds "very loud." Therefore it is convenient to append a scale of purely verbal description, such as "quiet," "very loud," etc., alongside a decibel scale of loudness. Figure 10.4 shows the decibel levels and verbal loudness descriptions of various sounds. The upper level is about 120 decibels or a little more; such sounds effectively saturate our sense of hearing, and greater sound powers are not perceived as being louder. Of course, they may be more harmful, but we are discussing only sensations. Note the extreme range of the scale: a 120-decibel sound is a trillion times more powerful than a zero-decibel sound.

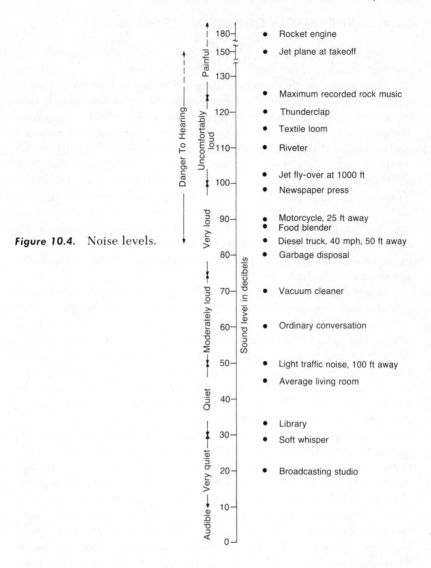

Figure 10.4. Noise levels.

Several related decibel scales have been proposed that are designed to reflect various annoyance factors. One such scale was developed specifically to rate aircraft noise. However, they all take the same logarithmic form shown earlier.

10.4   THE EFFECTS OF NOISE

Noise can interfere with our communication, diminish our hearing, and affect our health and our behavior.

Interference With Communication.   We have defined noise as unwanted sound. Let us think for a moment about the sounds that we do want: live or recorded speech or music, danger warnings such as the cry of a baby in a distant room or the rattle of a rattlesnake, pleasant natural sounds such as the chirp of a bird or the rustle of leaves in a gentle breeze. We want to hear these sounds at the right level — not too loud or too quiet — and without interference. Noise keeps us from what we want to hear, with the result that we do not hear it so well, or at all, or that the sound we want to receive must be unpleasantly loud for us to get the message. A conversation in the still forest mentioned at the beginning of this chapter can be carried on in whispers; but we must shout to each other to make sense in a boiler factory.

Loss of Hearing.   An occasional noise interferes with unwanted sounds, but we recover when quiet is restored. However, if the exposure to loud noise is protracted, some hearing loss becomes permanent. The general level of city noise, for example, is high enough to deafen us gradually as we grow older. In the absence of such noise, hearing ability need not deteriorate with advancing age. Thus, inhabitants of quiet societies, such as tribesmen in southeastern Sudan, hear as well in their seventies as New Yorkers do in their twenties.

In general, noise levels of about 80 decibels or higher can produce permanent hearing loss, although, of course, the effect is faster for louder noises, and it is somewhat dependent on the frequency. At a 2000-cycle frequency, for example, it is estimated that exposure to 95-decibel noise (about as loud as a power lawn mower) will depress one's hearing ability by about 15 decibels in 10 years. Occupational noise, such as that produced by bulldozers, jackhammers, diesel trucks, and aircraft, is deafening many millions of workers. The fact that women in technologically developed societies hear better than men is undoubtedly related to the fact that they are less exposed to occupational noise; in undeveloped, quiet societies, women and men hear equally well. This difference between the sexes in developed societies may diminish as more effort is exerted to make industry quiet, and as more women are exposed to noisy household appliances or to radio or television speakers turned up high so that they may be heard over the vacuum cleaner. Battle noises, such as those made by tanks, helicopters, jets, and artillery, are so deafening that more than half of the American soldiers who undergo combat training suffer enough hearing loss so that they could be considered physically unfit for combat.

Recent concern over exposure of people to rock music stems from the fact that such music is often indeed very loud. Sound levels of 125 decibels have been recorded in some discothèques. Such noise is at the edge of pain and is unquestionably deafening. Analogous to listening to deafening music would be staring at the sun, which causes blindness.

Other Effects on Health and Behavior. As we have already discussed in many contexts, a living organism, such as a human being, is a very complicated system, and the effects of a stress or a disturbance follow intricate pathways that may be very difficult to elucidate. Having read this book thus far, you should be skeptical if you are told that a disturbance great enough to deafen you will have no other effects. Indeed, many investigators believe that loss of hearing is not the most serious consequence of excess noise. The first effects are anxiety and stress reactions or, in extreme cases, fright. These reactions accompany a change in the hormone content of the blood, which in turn produces body changes such as increased rate of heart beat, constriction of blood vessels, digestive spasms, and dilation of the pupils of the eyes.* The long-term effects of such overstimulation are difficult to assess, but we do know that in animals it damages the heart, brain, and liver and produces emotional disturbances. The emotional effects on people are, of course, also difficult to measure. We do know that work efficiency goes down when noise goes up.

10.5 NOISE CONTROL

Noise is transmitted from a source to a receiver. To control noise, therefore, we can reduce the source, interrupt the path of transmission, or protect the receiver.

Reducing the Source. The most obvious source reduction is simply the reduction of the sound power. Don't beat the drum so hard, or ring the bell so loud, or run so many trucks or motorcycles, or mow the lawn with a power mower so often. There are obvious limitations to this type of solution; for example, if we run fewer trucks, we will have less food and other essentials delivered to us.

---

*Hormones are biochemical regulatory substances that act in trace amounts. They are produced by various glands in the body.

Even if we do not reduce the sound power, we may be able to reduce the noise production by changing the source in some way. Our purpose in pushing a squeaky baby carriage is to move the baby, not to make noise. Therefore, we can oil the wheels to reduce the squeaking. In addition, machinery should be designed so that parts do not needlessly hit or rub against each other.

It might be possible to modify technological approaches so as to accomplish given objectives more quietly. Rotary saws instead of jackhammers could be used to break up street pavement. Ultrasonic pile drivers could replace the noisier steam-powered impact-type pile drivers.

We could also change our procedures. If a city sidewalk must be broken up by jackhammers, it would be better not to start early in the morning, when many people are asleep, but later in the day, when many have left for work. Aircraft takeoffs could be preferentially routed over less densely inhabited areas.

Interrupting the Path. We have learned that sound travels through air by compressions and expansions. It also travels through other springy media, including solids such as wood. Such solids vibrate in response to sound and therefore do not effectively interrupt its transmission, as many residents of multiple dwellings will readily attest. However, we could use various materials that vibrate very inefficiently, such as wool, and absorb the sound energy, converting it to heat. (Very little heat is involved; the sound power of a symphony orchestra will warm up a room about as much as a ten-watt electric heater.) Sound-absorbing media have been developed extensively; they are called **acoustical materials**. We could also build interruption of the sound waves mechanically into more kinds of machinery; devices that function in this way are called **mufflers**. Finally, we may be able to deflect the sound path away from the receiver, as by mechanically directing jet exhaust noise upward instead of down. Such deflection is, in effect, an interruption between source and receiver.

Protecting the Receiver. We protect ourselves instinctively when we hold our hands over our ears. Alternatively, we can use ear plugs or muffs. (Stuffing in a bit of cotton does very little good.) A combination of ear plugs and muff can reduce noise by 40 or 50 decibels, which could make a jet plane sound no louder than a vacuum cleaner. Such protection could prevent the deafness caused by combat training, and should also be worn for recreational shooting.

We can also protect ourselves from a noise source by going away from it. In a factory, such reduction of exposure may take the form of rotating assignments so that different workers take their turns at the noisy jobs.

10.6   A PARTICULAR CASE:
THE SUPERSONIC TRANSPORT (SST)

The SST is a passenger aircraft that travels faster than sound and at much higher altitudes than subsonic airplanes. Higher speed requires more power, and more power makes more noise. Near airports, the noise problem is associated with takeoff and the rapid climb shortly after, although these speeds are subsonic. The engines on an SST must be small in diameter to provide optimal streamlining, and the noise from jet exhaust increases very rapidly (for a given engine thrust) as jet diameter is reduced and its speed increased. The statement has been made that the proposed American SST, on the runway, would sound like 50 ordinary jets taking off at the same time. This statement has engendered considerable controversy because it has been interpreted by some to mean that the SST would be 50 times as loud as an ordinary jet. We can understand the situation better by analogy to the earlier decibel formula:

$$\text{Loudness (compared with one jet plane)} = 10 \times \log_{10} \left(\frac{\text{sound power of 50 jets}}{\text{sound power of 1 jet}} \right)$$
$$= 10 \times \log_{10}50$$
$$= 10 \times 1.7 = 17 \text{ decibels}$$

(Note: The log of 50, as obtained from log tables, is 1.7)

Note that this scale is *not* the same as the one that is based on the least audible sound. This scale is based on the noise of one jet plane. (However, the same result would have been obtained if each aircraft had been compared separately with the least audible sound on the standard decibel scale. See Problem 9.) The calculations tell us that 50 jet planes or their equivalent with reference to sound (one SST) are 17 decibels louder than one jet plane. However, the calculation does not tell you how much more annoyance this would cause you; nor does it tell what you would do about it. Experiments show that a sound 17 decibels louder than another is judged "three to four times as loud," but it sounds nevertheless just like 50 of the weaker sounds all sounding simultaneously. Experiments on airport noise show that a single aircraft 10 decibels louder than another produces about the same annoyance as 10 separate flights of the quieter craft spread throughout the day.

This relationship is used by the government and by airport operators in planning land use around airports.

When the SST reaches supersonic speed in flight, another effect, the **sonic boom**, occurs. To visualize this effect, think of a speed boat moving rapidly in the water. Its speed is greater than that of the waves it creates, and it therefore leaves its waves behind it. Moreover, the wave energy is being continuously reinforced by the forward movement of the boat. The result is a high-energy wave, called a **wake**, that trails the boat in the shape of a V and that slaps hard against other vessels or against a shoreline. The sonic boom is a high-energy air wave of the same type. The tip of the wake moves forward with the airplane, while the sound itself moves out from the wake at its usual speed. The faster the airplane, the more slender is the wake.

To be struck unexpectedly by a sonic boom can be quite unnerving. It sounds like a loud, close thunderclap, which can seem quite eerie when it comes from a cloudless sky. The duration of the boom is only about $1/10$ to $1/2$ second, during which time the pressure rises above normal, then falls below normal. Depending on the power it generates, the sonic boom can rattle windows or shatter them or even destroy buildings. It is important to avoid the misconception that the sonic boom occurs only when the aircraft "breaks the sound barrier," that is, passes from subsonic to supersonic speed. On the contrary, the sonic boom is continuous and, like the wake of a speedboat, trails the aircraft all during the time that its speed is supersonic. Furthermore, the energy of the sonic boom increases as the supersonic speed of the aircraft increases.

Supersonic speeds are usually measured in mach numbers:*

$$\text{mach number} = \frac{\text{speed of object}}{\text{speed of sound}}$$

If an object is traveling at the speed of sound, then the numerator and the denominator of the equation are the same, and the mach number equals 1. Mach 2 is twice the speed of sound, mach 3 is three times the speed of sound, and so forth. The higher the mach number at a given altitude, the greater the energy and, hence, the destructive effect of the sonic boom.

The high altitude of some SST flight patterns (60,000 to 70,000 feet) poses another kind of problem—stratospheric air pollution. The SST discharges water, carbon dioxide, oxides of nitrogen, and

---

*After Ernst Mach, 1838–1916, a physicist who made important discoveries about sound.

particulate matter. The question is, can the effects of these pollutants be harmful? The answer, at this time, is speculative. For
example, it is estimated that a fleet of 500 SST's over a period
of years could increase the water content of the stratosphere by
50 to 100 per cent, which could result in a rise in average temperature of the surface of the Earth of perhaps 0.3°F and could cause
destruction of some of the stratospheric ozone that protects Earth
from ultraviolet radiation. We cannot predict with confidence
whether such effects would be trivial or serious. On one hand, we
can take the position that we should not trifle with the natural
condition of Earth in order to save a few hours of long-distance
travel time for a tiny fraction of the population. On the other, we
can say that supersonic travel will speed up progress throughout
the world by transporting technical experts rapidly to places where
they are needed, that it will create many new jobs, and that continued research will teach us how to lessen the pollution problems.
These, again, are the trade-offs.

But, stop. Why have we been discussing air pollution in a noise
chapter? You see, it's like this: The various parts of an ecosystem
are functionally related; what affects one affects all. This inter-
dependence, this difficulty in rigidly defining separate compart-
ments, applies not only to ecosystems, but even to books about
ecology.

PROBLEMS

1. Define elastic wave; wavelength; frequency; amplitude.

2. What is noise? Do you think it would be feasible to develop
   an instrument that would indicate how noisy a given sound is?
   Defend your answer.

3. Define or explain: mach number; sonic boom.

4. A man carries a decibel meter with him for a day and records
   the following readings in his diary:

7:00 A.M.	Baby crying.	84 dB
7:30	Dishwasher in kitchen.	70
7:45	Garbage truck, 150 feet away.	90
8:00	Traffic noise while waiting for bus.	81
8:45	Arrived at entrance to office. Noise of jackhammer on sidewalk.	106

9:00–12:00 Noon	Average sound in office.	45
12:00–1:00 P.M.	Noise in restaurant – dishes, etc.	45
5:00–5:30	Rode home on subway (windows open).	90–111
6:00	Mowed lawn with power mower.	93

Offer suggestions for reducing the perceived loudness of each of these various noises.

The following questions require arithmetic or algebraic reasoning or computation.

5. As stated in Section 10.3, the scale of subjective loudness is based on a pure tone of 1000 cycles per second. Calculate the wave length of this tone if the speed of the sound is 1100 feet per second.

6. A person in a living room makes a singing sound that has 10 thousand times the power of the faintest audible sound. What is his loudness in decibels?

7. The loudness of a motorcycle at a distance of 25 feet is 90 decibels. How many times more sound power than that of the faintest audible sound does this motorcycle have?

8. The statement is made in Section 10.3 that a 120-decibel sound is a trillion times more powerful than a zero-decibel sound. Show that this statement is true.

9. In Section 10.6 there is a calculation which shows that 50 jet planes, equivalent in sound to one SST, are 17 decibels louder than one jet plane. The formula used was based on the sound power of one jet, rather than the sound power of the least audible sound. Show that the same result could be obtained by using the formula based on the least audible sound. (You may assume that the loudness of one jet plane is 100 decibels. Calculate the ratio of this sound power to the faintest audible sound; then calculate the loudness in decibels of 50 times as much sound power, and subtract to get the difference.)

10. From the statements made in Section 10.6, how many *separate* takeoffs, spread throughout the day, of an ordinary jet would be as annoying, on the runway, as one SST? How many separate takeoffs of an ordinary jet would be as annoying on the runway as 50 ordinary jets all taking off at once?

11. A rocket is moving through the lower atmosphere, where the speed of sound is 1100 feet per second, at mach 2.5. What is the speed of the rocket in feet per second?

Answers

---

5. 1.1 ft

6. 40 dB

7. 1,000,000,000

8. One trillion $= 10^{12}$; $10 \log 10^{12} = 120$

9. Let x = sound power of 1 jet compared to least audible sound. Then loudness of 1 jet $= (10 \log x)$ dB, and loudness of 50 jets $= (10 \log 50 \text{ x}) \text{ dB} = (10 \log 50 + 10 \log x) \text{ dB} = (17 + 10 \log x) \text{ dB}$. Difference $= (17 + 10 \log x) - 10 \log x = 17$ decibels. The value of x can be assumed to be 100 dB or any other number; the 50 jets, or one SST, will still be 17 dB louder than 1 jet

10. 17; 17

11. 2750 ft/sec

BIBLIOGRAPHY

For a basic text on noise control, refer to

Leo L. Beranek. *Noise Reduction.* New York: McGraw-Hill Book Co., 1960. 752 pp.

Three recent popular books that take up the environmental aspects of noise are

Theodore Berland. *The Fight For Quiet.* New York: Prentice-Hall, 1970. 370 pp.
Robert Alex Baron. *The Tyranny of Noise.* New York: St. Martin's Press, 1970. 294 pp.
Henry Still. *In Quest of Quiet.* Harrisburg, Pa.: Stackpole Books, 1970. 220 pp.

APPENDIX

A.1   THE METRIC SYSTEM

Systems of measurement are used for the acquisition of knowledge of quantities in terms of standard units. The metric system, now used internationally in science, was originally established by international treaty at the Metric Convention in Paris in 1875 and has since been extended and improved. In this book, we are concerned with four of the fundamental metric units. These are:

Quantity	Unit	Abbreviation
length	meter	m
mass	kilogram	kg
time	second	sec
temperature	degree	°C

Larger or smaller units in the metric system are expressed by the following prefixes:

Multiple or Fraction	Prefix	Symbol
1000	kilo	k
1/100	centi	c
1/1000	milli	m
1/1,000,000	micro	μ

LENGTH

The **meter** was once defined in terms of the length of a standard bar; it is now defined in terms of wavelengths of light. A meter is about 1.1 yards.

1 **centimeter**, cm, is 1/100 of a meter, or about 0.4 inches.

The unit commonly used to express sizes of dust particles is the **micrometer**, μm, also called the **micron.** There are one million micrometers in a meter, or about 25,000 micrometers per inch.

MASS

The **kilogram**, kg, is the mass of a piece of platinum-iridium metal called the Prototype Kilogram Number 1, kept at the International Bureau of Weights and Measures, in France. It is equal to about 2.2 pounds.

One gram, g, is 1/1000 kg.

VOLUME

Volume is not a fundamental quantity; it is derived from length.
One **cubic centimeter**, cm³, is the volume of a cube whose edge is 1 cm.
One **liter** = 1000 cm³.

TEMPERATURE

If two bodies, A and B, are in contact, and if there is a spontaneous transfer of heat from A to B, then A is said to be *hotter* or at a higher temperature than B. Thus, the greater the tendency for heat to flow away from a body, the higher its temperature is.

The Celsius (formerly called Centigrade) temperature scale is defined by several fixed points. The most commonly used of these are the freezing point of water, 0°C, and the boiling point of water, 100°C.

The Fahrenheit scale, commonly used in medicine and engineering in England and the United States, designates the freezing point of water as 32°F and the boiling point of water as 212°F.

ENERGY

Energy is the capacity to do work. There is energy in a mule, in a moving train, in a compressed spring, in a stick of dynamite or a pound of coal or an ounce of uranium.

The metric unit of energy is the **joule**, which is defined in terms of physical work. The energy of 1 joule can lift a weight of 1 pound to a height of about 9 inches. An **erg** is a much smaller unit; there are 10 million ergs in a joule.

The unit commonly used to express heat energy, or the energies involved in chemical changes, is the **calorie**, cal. One calorie is about 4.2 joules. The energy of one calorie is sufficient to warm 1 gram of water 1°C.

The **kilocalorie**, kcal, is 1000 calories. This unit is also designated Calorie (capital C), especially when it is used to express food energies for nutrition.

A.2   CHEMICAL SYMBOLS, FORMULAS, AND EQUATIONS

Atoms or elements are denoted by symbols of one or two letters, like H, U, W, Ba, and Zn.

Compounds or molecules are represented by formulas that consist of symbols and subscripts, sometimes with parentheses. The subscript denotes the number of atoms of the element represented by the symbol to which it is attached. Thus H_2SO_4 is a formula that represents a molecule of sulfuric acid, or the substance sulfuric acid. The molecule consists of 2 atoms of hydrogen, 1 atom of sulfur, and 4 atoms of oxygen. The substance consists of matter that is an aggregate of such molecules. The formula for oxygen gas is O_2; this tells us that the molecules consist of 2 atoms each.

Chemical transformations are represented by chemical equations, which tell us the molecules or substances that react and the ones that are produced, and the molecular ratios of these reactions. The equation for the burning of methane in oxygen to produce carbon dioxide and water is:

$$CH_4 + 2O_2 \rightarrow CO_2 + 2H_2O$$

Each coefficient applies to the entire formula that follows it. Thus $2H_2O$ means $2(H_2O)$. This gives the the following molecular ratios: reacting materials, 2 molecules of oxygen to 1 of methane; products, 2 molecules of water to 1 of carbon dioxide. The above equation is balanced because the same number and kinds of atoms, one of carbon and four each of hydrogen and oxygen, appear on each side of the arrow.

The atoms in a molecule are held together by chemical bonds. Chemical bonds can be characterized by their length, the angles they make with other bonds, and their strength (that is, how much energy would be needed to break them apart).

$$O$$
$$\diagup \diagdown$$

The formula for water may be written as H   H, showing that the molecule contains two H-O bonds. The length of each bond is about 1/10,000 of a micrometer, and the angle between them is 105°. It would require about 12 kcal to break all of the bonds in a gram of water. These bonds are strong, as chemical bonds go.

In general, substances whose molecules have strong chemical bonds are stable, because it is energetically unprofitable to break strong bonds apart and rearrange the atoms to form other, weaker bonds. Therefore, stable substances may be regarded as chemically self-satisfied; they have little energy to offer, and are said to be energy-poor. Thus, water, with its strong H-O bonds, is not a fuel or a food. The bonds between carbon and oxygen in carbon dioxide, CO_2, are also strong (about 1.5 times as strong as the H-O bonds of water), and CO_2 is therefore also an energy-poor substance.

In contrast, the C-H bonds in methane, CH_4, are weaker than the H-O bonds of water. It is energetically profitable to break these bonds and produce the more stable ones in H_2O and CO_2. Methane is therefore an energy-rich substance and can be burned to heat houses and drive engines.

Table of Relative Atomic Weights (1969)

Based on the assigned relative atomic mass of $^{12}C = 12$

	Symbol	Atomic number	Atomic weight		Symbol	Atomic number	Atomic weight
Actinium	Ac	89		Mercury	Hg	80	200.59
Aluminum	Al	13	26.9815	Molybdenum	Mo	42	95.94
Americium	Am	95		Neodymium	Nd	60	144.24
Antimony	Sb	51	121.75	Neon	Ne	10	20.179
Argon	Ar	18	39.948	Neptunium	Np	93	237.0482
Arsenic	As	33	74.9216	Nickel	Ni	28	58.71
Astatine	At	85		Niobium	Nb	41	92.9064
Barium	Ba	56	137.34	Nitrogen	N	7	14.0067
Berkelium	Bk	97		Nobelium	No	102	
Beryllium	Be	4	9.01218	Osmium	Os	76	190.2
Bismuth	Bi	83	208.9806	Oxygen	O	8	15.9994
Boron	B	5	10.81	Palladium	Pd	46	106.4
Bromine	Br	35	79.904	Phosphorus	P	15	30.9738
Cadmium	Cd	48	112.40	Platinum	Pt	78	195.09
Calcium	Ca	20	40.08	Plutonium	Pu	94	
Californium	Cf	98		Polonium	Po	84	
Carbon	C	6	12.011	Potassium	K	19	39.102
Cerium	Ce	58	140.12	Praseodymium	Pr	59	140.9077
Cesium	Cs	55	132.9055	Promethium	Pm	61	
Chlorine	Cl	17	35.453	Protactinium	Pa	91	231.0359
Chromium	Cr	24	51.996	Radium	Ra	88	226.0254
Cobalt	Co	27	58.9332	Radon	Rn	86	
Copper	Cu	29	63.546	Rhenium	Re	75	186.2
Curium	Cm	96		Rhodium	Rh	45	102.9055
Dysprosium	Dy	66	162.50	Rubidium	Rb	37	85.4678
Einsteinium	Es	99		Ruthenium	Ru	44	101.07
Erbium	Er	68	167.26	Samarium	Sm	62	150.4
Europium	Eu	63	151.96	Scandium	Sc	21	44.9559
Fermium	Fm	100		Selenium	Se	34	78.96
Fluorine	F	9	18.9984	Silicon	Si	14	28.086
Francium	Fr	87		Silver	Ag	47	107.868
Gadolinium	Gd	64	157.25	Sodium	Na	11	22.9898
Gallium	Ga	31	69.72	Strontium	Sr	38	87.62
Germanium	Ge	32	72.59	Sulfur	S	16	32.06
Gold	Au	79	196.9665	Tantalum	Ta	73	180.9479
Hafnium	Hf	72	178.49	Technetium	Tc	43	98.9062
Helium	He	2	4.00260	Tellurium	Te	52	127.60
Holmium	Ho	67	164.9303	Terbium	Tb	65	158.9254
Hydrogen	H	1	1.0080	Thallium	Tl	81	204.37
Indium	In	49	114.82	Thorium	Th	90	232.0381
Iodine	I	53	126.9045	Thulium	Tm	69	168.9342
Iridium	Ir	77	192.22	Tin	Sn	50	118.69
Iron	Fe	26	55.847	Titanium	Ti	22	47.90
Krypton	Kr	36	83.80	Tungsten	W	74	183.85
Lanthanium	La	57	138.9055	Uranium	U	92	238.029
Lawrencium	Lr	103		Vanadium	V	23	50.9414
Lead	Pb	82	207.2	Wolfram	W	74	183.85
Lithium	Li	3	6.941	Xenon	Xe	54	131.30
Lutetium	Lu	71	174.97	Ytterbium	Yb	70	173.04
Magnesium	Mg	12	24.305	Yttrium	Y	39	88.9059
Manganese	Mn	25	54.9380	Zinc	Zn	30	65.37
Mendelevium	Md	101		Zirconium	Zr	40	91.22

INDEX